Ian D. Clark (Editor)

An Historical Geography of Tourism in Victoria, Australia – Case Studies

Ian D. Clark (Editor)

An Historical Geography of Tourism in Victoria, Australia – Case Studies

Managing Editor: Lucrezia Lopez

DE GRUYTER
OPEN

Published by De Gruyter Open Ltd, Warsaw/Berlin

ISBN: 978-3-11-037010-2
e-ISBN: 978-3-11-037011-9

Bibliographic information published by the Deutsche Nationalbibliothek
The Deutsche Nationalbibliothek lists this publication in the Deutsche Nationalbibliografie; detailed bibliographic data are available in the Internet at http://dnb.dnb.de.

Managing Editor: Lucrezia Lopez

www.degruyteropen.com

Cover illustration: *Holiday Excursionists on the You Yangs*

Contents

Abbreviations

AAP	Archaeological and Aboriginal Preservation
AAV	Aboriginal Affairs Victoria
AHU	Aboriginal Heritage Unit
AIAS	Australian Institute of Aboriginal Studies
BSC	Buchan Sesquicentenary Committee
CCL	Commissioner of Crown Lands
CFL	Conservation Forests & Lands
DCE	Department of Conservation & Environment
IRS	Infra-red Spectroscopy analysis
Jnl	Journal
LCC	Land Conservation Council of Victoria
ML	Mitchell Library, State Library of New South Wales
MOV	Museum of Victoria
n.d.	No Date
NMV	National Museum of Victoria
NRE	Department of Natural Resources and Environment
PROV	Public Records Office Victoria
PV	Parks Victoria
RHSV	Royal Historical Society of Victoria
SEM	Scanning Electron Microscopy analysis
SLV	State Library of Victoria
TAFE	Technical and Further Education
USEC	Urban Spatial and Economic Consultants Pty Ltd.
XRD	X-ray Diffraction analysis
XRF	X-ray Fluorescence analysis
VAS	Victoria Archaeological Survey
VDL	Van Diemen's Land
VPRS	Victorian Public Record Series

Metric Conversions

Throughout this book weights and measures have been expressed in contemporary terms: imperial before 1970 and metric thereafter.
The British Pound and, after 1900, the Australian Pound, was the basic unit of currency in Victoria until the change to decimal currency in 1966 when one Pound was worth two Dollars. Contemporary monetary units have been expressed in this work. Before 1966, these were Pounds, Shillings, and Pence. The following conversion factors apply:

1 inch (in) = 0.0254 m
1 foot (ft) = 0.305 m
1 yard (yd) = 0.914 m

1 rod = 5½ yards = 16½ feet = 5.0292 m
1 mile = 1,609.4 m

1 ounce (oz) = 28.3 g
1 pound (lb) = 454 g
1 ton = 1.02 tonne

1 bushel (bus) = 0.0364 cubic metres

1 rood = ¼ acre
1 perch = 30¼ square yards = 25.29 sq. metres
1 acre = 0.405 ha

1 penny (1d) = 0.83 cents
12 pence = 1 shilling
1 shilling = 10 cents
1 pound (£) = 2 dollars

Acknowledgements

We would like to thank the following people and organizations for their assistance during this study: staff at the National Gallery of Australia, Canberra; National Gallery of Victoria, Melbourne; Warrnambool Art Gallery; National Library Australia, Canberra; Department of Environment and Primary Industries, Victoria; and Clare Gervasoni, Curator, Geoffrey Blainey Collection, Federation University Australia, for assistance with image reproduction and permissions. Special thanks to Peter Clark and Emily Slattery for permission to reproduce their photographs of Bunjils Shelter and Hanging Rock, respectively. Finally, special thanks go to Managing Editor, Lucrezia Lopez, and all the staff at De Gruyter Open who were associated with this work for their outstanding assistance.

Biographical Notes

Dr Ian D. Clark is a Professor of Tourism in the Faculty of Business at Federation University Australia. He has a Doctorate in Aboriginal Historical Geography from Monash University. His areas of interest include the history of tourism, cultural heritage management, attractions management, Indigenous tourism, Victorian Aboriginal history, and placenames.

Ever Dolce has a Bachelor of Commerce degree from the University of Ballarat. His areas of interest include colonial history, environmental sustainability, macroeconomics and social justice.

Lisa Justin has a Bachelor of Management degree from Federation University Australia. Her area of research interest is tourism.

Sharnee Sergi has a Bachelor of Rural Social Welfare degree from Federation University Australia. Her professional areas of interest include community development, social justice and human rights, and an ongoing commitment to research and building evidence-based practice.

Stephanie Skidmore has a Bachelor of Tourism Management degree from Federation University Australia. Her interests include local history as well as Indigenous history of Australia.

Jaimie Watson has a Bachelor of Arts Humanities and Social Sciences degree from the University of Ballarat. Her areas of interest include 20th century history, Australian history and sociology.

1 Introduction: Nascent Tourism in Victoria, Australia – Insights Into the Evolution of Its Tourism Landscape by Ian D. Clark

This work is concerned with the emergence of tourism in colonial Victoria, Australia, and is part of ongoing research into understanding Victoria's 'tourism era of discovery' (Towner, 1996: 140). It is concerned with the processes of 'opening up' new attractions and its focus is the discovery state of the development of tourism or what Young (1983) has termed 'pretourism'. Victoria's tourism era of discovery, here defined as 'nascent tourism' or 'pretourism', is a period that has generally been neglected in tourism histories in Australia, notwithstanding the recent works of Bonyhady (2000), Horne (2005), and Inglis (2007). Nascent tourism, defined as the embryonic or emergent phase in which natural attractions are coming into being as the subject of tourist visitation, will be contextualized in the study of eight tourism sites that will be the primary focus of this work.

Travellers' accounts and other contemporary sources will be used to provide us with insights into Victoria's nascent tourism – through them; we should be able to see the various places that were emerging as tourist sites in the colonial space. The sources are interrogated as journals or narratives that offer a biography of the journey in ways similar to Carter's (1988) and Ryan's (1996) interrogations of the journals of Australian explorers. These accounts enable observations of tourism and travel phenomena to be contrasted and allow geographical and temporal controls to be applied. Accounts from the 1830s and 1840s, for example, capture the nascent state of hospitality and travel as it was centred around squatting stations; the 1850s and 1860s show the evolution of an accommodation industry away from Melbourne and the improvement of transport infrastructure contrasted with the chaos caused by the gold rushes and the emergence of fledgling townships such as Ararat and Ballarat.

1.1 Understanding the Colonial Settler Gaze

To understand the colonial settler gaze in Victoria, it is necessary to understand the conventions or tropes that mediated it. The cultural milieu of the various travellers needs to be contextualized, especially the prevailing paradigms or conventions of seeing, particularly the 'cult of the Sublime', the 'cult of the Gothic' (Ousby, 2002), the picturesque and the panoramic (Ryan, 1996). Renderings of the new world landscape in terms of old world paradigms, or the notion of pictorial colonization, should emerge in the various travel accounts. Furthermore, the role of settlers in mediating tourism, something I dubbed 'nascent private tourism' will be scrutinized – especially the scenic attractions and other places of interest that settlers, as hosts, showed their guests. In these settings, the settlers were themselves discoverers and explorers.

Henry Brown's (1862: 35) experience upon disembarking in Melbourne encapsulates the 'shock of the new' that confronted many travellers: 'I can truly say that I left the ship with a sigh of regret. It may be that the strangeness of all around made me cling instinctively to something that had been in England, and to which I was accustomed, but I have since learnt that there are few who leave a vessel, where they have been comfortable, without similar feelings'. Richard Howitt (1845: 168) also discussed the problems emigrants faced in new colonies, something he described as 'home-return-anxiety', where emigrants are personally 'abroad, but mentally at home; living, moving, and having their existence amongst friends and kindred'. Kinahan Cornwallis (1859: 33f) confirmed that first impressions of Australia were often incorrect: 'The beauties of Australia have been frequently painted in the brightest and most inviting of hues, and I had read those pleasant book pictures before my embarkation from England; but instead of the beautiful, I had as yet only experienced the wretched, and on this, my first night in Australia ... After experience however proved to me that Australia abounded less in shadow than in sunshine, and that my first experiences of the country were the worst'. Richard Twopeny (1870: 1) counselled his readers that 'In one sense the visitor is disappointed with his first day in an Australian city. The novelties and differences from the Old Country do not strike him nearly so much as the resemblances. It is only as he gets to know the place better that he begins to notice the differences. The first prevailing impression is that a slice of Liverpool has been bodily transplanted to the Antipodes, that you must have landed in England again by mistake, and it is only by degrees that you begin to see that the resemblance is more superficial than real'.

Several travellers considered Victoria to be very English. Frenchman, Ludovic Marquis de Beauvoir (1870: 19, 30), for example, was struck by the Englishness of Victoria and although he believed it was 'a good thing to arrive at a place without any preconceived notions or prejudices, to wait for and seize upon first impressions, though very likely riper experience may change one's opinion', 'since I have landed it has struck me that the local tone of the country consists precisely in being no local tone at all, and that the colony, contrary to custom, resembles the mother country in a very unusual manner'. Samuel Smiles (1880: 179) who spent 18 months working as an accountant at Majorca, near Maryborough, during 1868-9, considered life in Victoria was very much like life in England. 'There are the same people, the same callings, the same pleasures and pursuits ... Indeed, Victoria is only another England, with a difference, at the Antipodes. The character, the habits of life, and tone of thought of the people, are essentially English'. Charles Carter (1870: 188) was struck with the English appearance of the country on both sides of the railway, at Malmesbury [sic], Kyneton, Woodend, all about Mount Macedon', and Mossman and Banister (1853: 62) considered the 'open forest-lands' of Victoria 'have very much the appearance of Hyde Park and Kensington Gardens'.

But Clara Aspinall (1862: 162) did not share their opinion and considered that Australian scenery 'cannot certainly bear comparison with British scenery; at the same time, I must add that I have seen some very pretty spots, and some very fine views in Australia; but whenever this happened to me, my first exclamation always was, "How very lovely! How very English!" Generally speaking, there is a monotony in the scenery of Australia which is wearisome to the eye. ... There is one want in Australia, which must always be felt in a new country by the traveler who is in search of the picturesque: namely, the want of scenes, ruins, or edifices hallowed by a sense of antiquity'. Robert Henderson (1911), an evangelist in Australia, concurred with Aspinall: 'Face to face with a new country and new conditions, I soon saw that, with the exception of a few centres of population and scattered villages, a mere handful of people were in possession of a vast Island-Continent, consisting of great flock masters and their shepherds, of farmers and gold-diggers, who had taken over the hunting grounds of the Aborigines, and were driving them into the interior and decimating them. In vain you looked for venerable cathedrals, baronial halls, old castles, famous battlefields, Druidical remains, or ancient history. Everything was new, and in the interior, wild and primitive. The only thing that could boast of antiquity was the black man, the native of the soil' (Henderson, 1911: 149). However, Christopher Hodgson (1846: 174) whose interests were botanical and geological, did not share Aspinall's or Henderson's opinions, and considered that 'Australia to the geologist is a truly interesting and wonderful country; unfolding new mysteries every day, and leaving simple man to revel in the midst of wonder, uncertainty and amazement'.

To understand the history of tourism visitation and the evolution of tourism attractions, this study uses perspectives developed by MacCannell (1976), Butler (1980), and Gunn (1994). MacCannell's (1976) research into the development of secular attractions through five stages – sight sacralization or naming, framing and elevation, enshrinement, duplication, and social reproduction will be tested to see if it satisfactorily accounts for the development of the eight attractions that are the focus of this study. Butler's (1980) tourism area life-cycle model may explain any subsequent stagnation and decline of the attractions. Finally, Gunn's (1994) spatial model of attractions should be able to add a spatial dimension to understanding the history of recreation planning at each attraction in terms of the three zones (nucleus, inviolate belt, and zone of closure) of visitor interaction outlined in his model.

MacCannell's model has been applied to specific studies of particular tourism sites, including Jacobsen's (1997) study of a cape in Norway; Clark's (2002b) preliminary study of Lal Lal Falls in Victoria, Australia, Slade's (2003) investigation of Gallipoli, Turkey, and more recently Forristal, Marsh, and Lehto's (2011) analysis of Historic Prophetstown in Battle Ground, Indiana, in the United States. Forristal *et al* (2011: 574) note that the 'use of MacCannell's site sacralisation model is not a quantitative exercise, but rather a subjective and qualitative one. There is no specific or agreed-upon operationalization of the model. Indeed, the beauty and strength of the model lies in its flexible application to a wide range of sites and situations'.

1.2 First Phase: Sight Sacralization and Naming

MacCannell (1976) has identified the first phase in the development of attractions as 'sight sacralization' or 'naming' when the sight or site is given a name. A fundamental step in its demarcation as a place of interest is its naming. In the case of Gunn's (1994) spatial model of attractions, the site is the nucleus of the attraction, the principal focus of visitor interest. This sacralization stage corresponds with Butler's (1980) 'exploration' stage in which tourism as such is nascent and visitor numbers are dispersed and insignificant, a stage best described as 'pre-tourism'.

1.3 Second Phase: Framing and Elevation

The second phase identified by MacCannell (1976) in the evolution of attractions is 'framing and elevation' which he argued results from an increase in visitation, when demand requires some form of management intervention, whereby the sight is displayed more prominently and framed off. MacCannell's framing and elevation phase corresponds with Gunn's (1994) 'inviolate belt' zone in his spatial model. The inviolate belt represents the essential setting of the nucleus, it is the area immediately surrounding it, and serves as the psychological setting for introducing the visitor to the attraction. This period also correlates with Butler's 'involvement' stage, which is characterised by an increase in tourist visitation and the emergence of an incipient tourism industry developing around the destination.

1.4 Third Phase: Enshrinement

MacCannell (1976) has identified 'enshrinement' as the third phase in the evolution of attractions. By this he means a phase where there is an increase in tourism numbers and the attraction's reputation is enhanced. During this phase the attraction developed its third and final spatial zone, as outlined by Gunn (1994) – that of its 'zone of closure', an outer area of community influence of travel structures such as land uses for modern travel services, such as railway stations and grandstands. The enshrinement phase equates with Butler's (1980) 'development', 'consolidation', and 'stagnation' stages, which are characterised initially by rapid tourism growth and dramatic changes in the tourism industry associated with the destination, its consolidation, and then its stagnation represented by declining visitor numbers.

1.5 Fourth Phase: Duplication

MacCannell's (1976) fourth phase in the development of attractions, is that of 'duplication', when copies of the nucleus of the attraction, are made available through media such as paintings, photographs, and postcards.

1.6 Fifth Phase: Social Reproduction

For MacCannell, the final stage is social reproduction which 'occurs when groups, cities, and regions begin to name themselves after famous attractions'.

Through a series of case studies from Victoria, summarized below, this work complements Horne's (2005) exploration of the evolution of ideas of wonder in scenic Australia in the nineteenth century that helped to create both a tourism industry and an enduring interest in the natural environment. For Horne (2005: 8), the 'answer to the question of why particular natural features became favourite destinations for nineteenth-century tourists in Australia is in their appeal to certain nineteenth-century cultural interests and sensibilities, the wonder expected to be inspired by a view, a geological formation, a botanical specimen'. Indeed, the sites that are studied in this book are themselves vestiges of nineteenth and early twentieth century sensibilities – distinctive scenery such as waterfalls, rocky outcrops, extinct volcanoes offering panoramic views, and limestone caves filled with exotic stalagmites and stalactites.

1.7 The Cases – Lal Lal Falls Scenic Reserve

The Lal Lal Falls, situated within the traditional country of the Wathawurrung Aboriginal people, is one of Victoria's most significant Aboriginal cultural sites, as it is one of several recorded living sites of Bunjil – the Kulin peoples' creator spirit. However, the Lal Lal Falls became a tourist attraction more for its natural significance than its Indigenous cultural values. Its reservation as a public reserve in 1865 marked the emergence of the site as an attraction with a nucleus and an essential setting. Tourism at Lal Lal Falls experienced its zenith when it was associated with the annual Lal Lal Races and, with their cessation in 1938, the Lal Lal Falls Scenic Reserve, in terms of visitor numbers, has experienced continued stagnation and general decline; though this has now been arrested by major intervention works in 2004-2008. Since 1938, there have been periods of management activity when site works have been undertaken, particularly in 1963, 1980, and 2004-2008; however, the benefits of the first two actions appear to have been short-lived and the infrastructure associated with the 'inviolate belt' – visitor amenities such as toilets, fireplaces, seats and tables – were vandalised and allowed to fall into disrepair.

What is particularly poignant about the Lal Lal Falls is its sacredness in Indigenous value systems. Its significance cannot be overstated. Yet, although the Aboriginal significance of the site has been understated in previous site promotion and off-site interpretation, visitor information and travel guides on the Internet are beginning to focus on the site's Indigenous values. In many respects the landslip in 1990 and the two resultant fatalities and the subsequent management decision to attempt to restrict visits to the base of the falls represents the lowest point in the tourism history of the attraction.

Lal Lal Falls, as a tourism attraction – one with significant natural and cultural values – has undergone a fundamental transition from being 'the top tourist attraction' and 'one of the most significant heritage sites' in the Buninyong region, where thousands of people would congregate for picnic races, to an attraction 'virtually forgotten' in 2001, only to reemerge in 2012 to be the leading natural attraction in the Ballarat district.

1.8 Buchan Caves Reserve

Although, the presence of caves in the Buchan district was first mentioned in 1840, the first known visit from staff from the Surveyor-General's department did not take place until 1854. Nascent tourism began to occur from the 1870s. In 1889 the first systematic geological survey was undertaken and recommendations made to develop the site into a tourist attraction similar to the Jenolan Caves in New South Wales replete with a caretaker and the installation of electricity. Further recommendations by a geologist with the Mines Department in 1900 saw the reservation of some 65 hectares of Crown Land near Buchan by the Department of Crown Lands and Survey in 1901, and an additional 48 hectares in 1902. The camping reserve at Buchan Caves was proclaimed in 1930 and in 1938 the Buchan Caves National Park was officially opened.

The two show caves at Buchan, Fairy Cave and Royal Cave, were 'discovered' in 1907 and 1910 respectively. The opening to the Fairy Cave was enlarged using some gelignite and after the construction of pathways and wire netting to protect the stalagmites and stalactites, the cave was opened to the public in late 1907. Royal Cave was opened in late 1913 after reserve employees cut through a solid block of marble and used a large quantity of explosives to blast through 150 feet of rock. In 1920 a generating plant was installed at the complex which remained in use until 1969 when the site was connected to the state electricity grid. A notable feature of the Buchan Caves reserve is that it has been modelled on the United States of America National Parks Service, especially its adoption of 'parkitecture' styles for some of its buildings.

In terms of the agencies responsible for the protection and development of tourism at the site, it is possible to identify a mixture of government departments and geologists, the local Buchan community, regional progress association, and key individuals such as Frank Moon. The Buchan Caves have significant Aboriginal values, especially the widespread heritage of caves as places where the Nargun and wicked and mischievous Nyols lived. For this reason Aboriginal people were not in the habit of venturing deep into the limestone caverns.

1.9 Bunjils Shelter, Black Range Scenic Reserve

Bunjils Shelter did not become a tourist attraction until its public 'discovery' in 1957. Given its importance and fragility as an Aboriginal art site it has progressed through MacCannell's attraction development phases within a relatively short period of time. The framing and elevation phase at Bunjils Shelter occurred before there was a significant increase in visitation; and rather than have tourist demand force management intervention, it was the fragility and rarity of the site that demanded that relevant authorities protect the art, or what Gunn understands as the 'nucleus' of the attraction. The management interventions taken to protect the art such as drip lines and protective grilles equate with Gunn's 'inviolate belt' zone. The enshrinement phase at Bunjils Shelter is not a significant phenomenon, though it may be argued that this occurs when tourists find the pathway to the site and the various interpretive signs that are placed *en route* of interest in their own right. These interventions equate with Gunn's 'zone of closure'. The duplication phase has occurred in that a facsimile of the shelter was produced and displayed in a nearby theme park attraction in Stawell in 1975 and the art featured in an Australia Post postage stamp series in 1984 dedicated to the First Australians. Social reproduction of the attraction has been minimal probably due to the fact that it is not deemed appropriate for entities to name themselves after an Aboriginal art site, especially one that features a representation of the creator spirit Bunjil.

1.10 Den of Nargun, Mitchell River National Park

The Den of Nargun, set in the Mitchell River National Park in Gippsland, approximately 50km from Bairnsdale, has had a successful tourism history since its European discovery in 1875. 'Discovered' by Europeans in 1875, the Den of Nargun did not begin to experience significant numbers of visitors until the 1930s largely through the agency of bushwalkers and field naturalists. Although the site was declared a 'sanctuary' in 1938, formal site protection did not take place until November 1963 when the private company, Australian Paper Manufacturers Ltd, donated 163 acres that included the cave that saw the formation of the Glenaladale National Park. In 1986 the Glenaladale National Park was merged with adjoining land and renamed the Mitchell River National Park. Today, the site is managed by Parks Victoria and the Gunaikurnai Land and Waters Corporation.

1.11 Hanging Rock Recreation Reserve

Mt Diogenes, or Hanging Rock as it is more popularly known, was the subject of geological research by William von Blandowski in 1855; however it was not until Boxing Day 1864 that we have the first recorded account of the rock being used for leisure purposes when a picnic was organized by a group of tradesmen from nearby Kyneton. In 1869 entrepreneur William Adams purchased the Rock and set about developing the site into a pleasure resort. In 1866 he had purchased a half-acre block near the rock and built the 'Hanging Rock Hotel'. In 1869 Adams placed a weir across Five Mile Creek to form a lake suitable for small rowing boats and other small water craft, and with the help of neighbouring publicans he developed a full programme of sports. Over the next seven years Adams cleared a track for bush picnic horse races and built a pavilion for dancing and continued to cater for picnics and sporting carnivals. In 1876 Adams sold the Rock to a Melbourne warehouse owner who had no interest in promoting a tourist attraction so the new owner leased the grounds to the owners of the Hanging Rock Hotel.

Around this time community agitation began to stir for the Government to purchase Hanging Rock for the public. The community of Newham urged its shire council to lobby the Government. In 1884, after a series of negotiations the Government purchased the Rock (and added it to the existing Crown Land reserve of 96 acres, west of Hanging Rock, which had never been privately owned), formally gazetting the Hanging Rock Recreation Reserve in the same year. The reserve was placed under the management of the Newham Shire Council. In 1885 a Racing Club was organized. The Council formed a committee of management from within its own councillors and oversaw site maintenance, pest control, and necessary site improvements. The races at Easter and New Year's Day were significant events, for example, in 1911 some 20,000 attended the News Years' Day races. In the 1920s a caretaker's cottage was built and refreshments were provided by the caretaker's wife to those who visited the reserve.

In 1967 author Joan Lindsay published her novel 'Picnic at Hanging Rock' which was produced into a movie in 1975. The movie saw a substantial increase in international and domestic tourism to the Rock. In 1978 the Victorian Government made funds available for site improvements including barbeques, toilets, and car parks to cope with the increase of tourism that had resulted from the release of the film. In 1990 management of the reserve reverted to the local council. In 1993 the reserve was expanded with the purchase of an adjacent property of 22 ha. – it is now 88 ha in size.

In terms of the agency responsible for the emergence and development of tourism at Hanging Rock this case study differs from the others in this book for the early history of the Rock is tied up with the fact that it was alienated land and owned by an entrepreneur who could see the tourism possibilities the rock offered. From 1869 until 1884 the entrepreneur developed the site into a tourist resort. It returned to public land in 1884 only after community agitation.

1.12 Mount Buninyong Scenic Reserve

As a tourist attraction, Mount Buninyong–with its natural and cultural values, has transitioned from being a resource for farming purposes in colonial days and a scenic attraction for picnics into a highly recognizable public reserve that is central in the local primary school's environmental and Indigenous curriculum. Mount Buninyong is an important site for the acknowledgment of the ongoing connection of Aboriginal people with the land through creation stories and cultural sites and is an ideal site for interpreting Aboriginal cultural history.

Mount Buninyong is an area of high geological and geomorphological significance. It has excellent interpretive value as it is the most obvious example in Victoria of a breached scoria cone and associated lava flows. The site is easily accessible as it is the only major scoria cone with a deep crater that remains on public land in Victoria. The Mount is still used as a scenic attraction today, and the bond between the local community and Mount Buninyong has always been strong. This has been expressed through organized interest groups as well as in the varied recreational pursuits of individual residents. In relation to Butler's life cycle model, the Mount could be said to be at the 'rejuvenation' stage as it is continuously being maintained and upgraded. In early 2012, the *Ballarat Courier* undertook a survey of the Ballarat region's favourite natural attractions. Lal Lal Falls was considered the region's best, securing 37.5 per cent of the vote; Lake Wendouree came in second with 32.3 per cent, and Mt. Buninyong, third, with 9.4 per cent.

1.13 Tower Hill State Game Reserve

Tower Hill as a tourism attraction has significant natural values for local and international visitors. It became a tourist attraction more for its natural significance rather than its Indigenous values, although visitors to the site are able to learn about the Indigenous people and their way of life. Tower Hill was one of the first natural features in Victoria to be given a European name 'Peak of Reconnaissance' in March 1802 by the French explorers Baudin and Peron. A nested caldera, it is one of the most recently active volcanoes in western Victoria, and as well as these significant natural values it has significant Aboriginal heritage values as represented in the detailed Aboriginal microtoponymy that has been recorded in the ethnographic record. The name Tower Hill is descriptive and is believed to date from 1838. Although European settlement and land use began from this time, portions of the caldera remained as Crown Land and in 1866 local residents were successful in their endeavours to reserve Tower Hill when the Victorian Government declared it the Tower Hill Acclimatisation Reserve.

When schools inspector James Bonwick visited the site in 1857 he considered it sublime and grand and urged its reservation. Two years previously, local settler

James Dawson had commissioned German landscape painter Eugen von Guerard to paint Tower Hill. Years later, in 1961, this example of duplication of the site featured prominently in attempts to restore the vegetation of the island to that depicted in the von Guerard painting. In 1892, Tower Hill became Victoria's first National Park and was maintained as a park for public recreational purposes by the local shire council. However, continued environmental degradation of Tower Hill led to its omission from the 1956 National Parks Act. Local Field Naturalists were responsible for efforts to arrest the degradation and take active steps towards its environmental rejuvenation. In 1960 Tower Hill became Victoria's seventh Game Reserve, and the government department responsible, the Department of Fisheries and Wildlife set about establishing a natural history centre and embarked on a restoration program based on von Guerard's painting.

1.14 You Yangs Regional Park

The You Yangs have been shown to be a different entity to the sites that are the subject of the other case studies in this volume. It is not a single waterfall or a single mountain but rather a range of granitic hills, of which approximately only a third have been formed into the You Yangs Regional Park. The largest peak in the You Yangs is one of the earliest sites in Victoria that were visited by European – in this case Matthew Flinders in 1802.

In 1835 part of the You Yangs was retained as a Common and used for cattle grazing and timber cutting. This became a Crown Lands Timber Reserve of about 1,000 ha in 1866. In 1875 the reserve was increased to its present size of 2,000 ha. In 1958 some 355 ha were declared a scenic reserve, and in 1962 an additional 110 ha was given special purpose reserve status for the protection of birds. In 1964 the reserve became a state forest. In 1981 some 1,515 ha were added bringing the park to 2,000 ha. In 1992, the reserve became the You Yangs Regional Park –to provide opportunities for recreation and to protect and conserve natural ecosystems to the extent that this was consistent with its recreational role.

It has been a site for botanical research since 1853 when the government botanist von Mueller conducted field surveys there, and with the formation of the Field Naturalists Club of Victoria in 1880, regular excursions have been conducted to the reserve. In 1853, the You Yangs was not the subject of much visitation, and the *Geelong Advertiser* speculated that perhaps as few as 50 people had stood on its highest peak. The advent of the railway from Geelong to Little River saw the You Yangs become a favourite destination for day excursionists from Melbourne and Geelong.

The You Yangs has been and remains an important historical attraction, especially for people who are interested in the early colonial period when Europeans first explored and settled in Victoria. There is a wide range of diverse activities available at the park. The recent surge in mountain bike rider participation attests to the successful efforts of park management in rejuvenating visitor interest. The changing demographic demand for the park and the continued involvement of the two Wathawurrung organisations allows for an expanded offering of Wathawurrung cultural history.

1.15 Conclusion

The question of how a tourist site is created from what was effectively a blank tourism landscape requires further study for, as this study of visitation at eight significant sites has shown, it is too simplistic to attribute the development of tourism at a particular site to the agency of local landholders. In the case of Lal Lal Falls, the locus of causality of its evolution into a mature attraction, or what MacCannell has understood as the framing and elevation stage of his evolutionary model, rested with the local community led by a local newspaper, *The Star*, which sought to ensure the attraction did not become private property in 1856 when the State Government was considering its sale. Its permanent reservation in 1865 as a scenic reserve ensured that it would become the focus of significant tourism. Nascent tourism began at the Buchan caves from the 1870s, however in terms of the agencies responsible for the protection and development of tourism at the site, it is possible to identify a mixture of government departments and geologists, the local Buchan community, regional progress association, and key inviduals such as Frank Moon.

Bunjils Shelter did not become a tourist attraction until it was 'discovered' in 1957. Given its importance and fragility as an Aboriginal art site it has progressed through MacCannell's attraction development phases within a relatively short period. The framing and elevation phase at Bunjils Shelter occurred before there was a significant increase in visitation; and rather than have tourist demand force management intervention, it was the fragility and rarity of the site that demanded that relevant government authorities protect the art, or what Gunn understands as the 'nucleus' of the attraction.

Although 'discovered' by Europeans in 1875, the Den of Nargun did not begin to experience significant numbers of visitors until the 1930s largely through the agency of bushwalkers and field naturalists. Although the site was declared a 'sanctuary' in 1938, formal site protection did not take place until November 1963 when the private company, Australian Paper Manufacturers Ltd, donated 163 acres that included the cave that saw the formation of the Glenaladale National Park.

In terms of the agency responsible for the emergence and development of tourism at Hanging Rock this case study differs from the others in this book for the early history of the Rock is tied up with the fact that it was alienated land owned by an entrepreneur who could see the tourism possibilities the rock offered. From 1869 until 1884 the entrepreneur developed the site into a tourist resort. It returned to public land in 1884 only after community agitation.

Mount Buninyong was first subject to European visitation in 1837, although it was first declared a public reserve in 1866 largely through the agency of the local municipal council.

Tower Hill was first named by Europeans in 1802, and a small Crown Land allotment at the site remained in public hands and was declared a reserve for public purposes in 1866. In 1892 it became Victoria's first National Park, and in 1961 it became a State Game Reserve.

The You Yangs, were also first named and visited by Europeans in 1802. In 1835 when Crown Lands were being made available for alienation, a section of land was retained as a Common for cattle-grazing and timber cutting. This Common became a Timber Reserve in 1866 and eventually became the You Yangs Regional Park in 1992.

The Buchan Caves and the Den of Nargun need to be contextualized in relation to their place in the pursuit of wonder in the nineteenth and early twentieth century. Horne (2005: 250) has shown how limestone caves were seen 'as representing the concepts of the sublime and the beautiful'. 'Their commercial development helped to open this experience to travellers, but the strong government involvement in financing them was a form of colonial altruism, a public investment in providing opportunities for the edification of the people through inspiration in response to the wonders of nature' (Horne, 2005: 251). Similarly, mountains such as the You Yangs, Hanging Rock, Mt Buninyong, and Tower Hill, and the Lal Lal Falls, in the mid-nineteenth century were understood in the language of the sublime.

All eight places that are the focus of this book have significant Indigenous values – in the case of Lal Lal Falls and Bunjils Shelter they are directly related to Bunjil – the creator spirit. In the case of Buchan Caves, Den of Nargun and Mt Buninyong we have Aboriginal stories that explain the significance of the sites – with Buchan, we have Aboriginal accounts of how caves were the home of evil beings such as Nyols and in the case of Den of Nargun, it was the home of a terrible stone-being called a Nargun; Buninyong was formed when two ancestral beings Buninyong and Derrinallum fought one another. The Aboriginal significance of Hanging Rock, You Yangs, and Tower Hill is demonstrated by accounts of Aboriginal association, Aboriginal place names, and cultural sites.

This work is part of a larger research project into the history of tourism in colonial Victoria, Australia, that is concerned with a fundamental set of questions: how does a tourist site come in to being? How does a tourist gaze emerge in a 'settler society'? How does an 'era of discovery' segue into 'tourism'? And, how was the tourist map of Victoria created by settler colonists? The tourist gaze in colonial Victoria was essentially mediated by Old World paradigms such as the picturesque and the panoramic. These sensibilities were shaping the gaze of British colonists and travellers and the Victorian landscape was seen through Old World lens. Although they share common Indigenous values, the application of classical models of attraction evolution to the eight sites that are the focus of this study, has shown that their evolution into natural and cultural tourism attractions is diverse and that Victoria's tourism landscape is dynamic and constantly changing. Some sites, for example, were reserved in the nineteenth century but others did not become tourism attractions until the twentieth century – such as the Den of Nargun in 1938, and Bunjils Shelter as recently as 1957. There are many other significant natural and cultural attractions in Victoria and much more research needs to be undertaken to understand more fully the evolution of Victoria's tourism landscape.

References

Aspinall, C. (1862). *Three years in Melbourne*. London: L Booth.
Bonyhady, T. (2000). *The Colonial Earth*. Melbourne: Melbourne University Press.
Brown, H. (1862). *Victoria, as I found it, during five years of adventure*. London: T. Cautley Newby.
Butler, R.W. (1980). The concept of a tourist area cycle of evolution: Implications for management of resources. *The Canadian Geographer*, 24(1), 5-12.
Carter, C.R. (1870). *Victoria, the British 'El Dorado'....* London: Edward Stanford.
Carter, P. (1988). *The Road to Botany Bay: An Essay in Spatial History*. London: Faber & Faber.
Clark, I.D. (2002a). Rock Art Sites in Victoria, Australia: a management history framework. *Tourism Management*, 23, 455-464.
Clark, I.D. (2002b). The ebb and flow of tourism at Lal Lal Falls, Victoria: a tourism history of a sacred Aboriginal site. *Australian Aboriginal Studies*, No. 2, 45-53.
Clark, I.D. (2009). Naming sites: names as management tools in Indigenous tourism sites – an Australian case study, *Tourism Management*, 30 (1), 109-111.
Clark, I.D. (2010). Colonial tourism in Victoria, Australia, in the 1840s: George Augustus Robinson as a Nascent Tourist. *International Journal of Tourism Research*, 12, (5), 561-573.
Cornwallis, K. (1859). *A panorama of the New World*. London: TC Newby.
De Beauvoir, L.M. (1870). *A voyage round the world*. London: John Murray.
Forristal, L.J., Marsh, D.G. & Lehto, X.Y. (2011). Revisiting MacCannell's Site Sacralization Theory as an Analytical Tool: Historic Prophetstown as a Case Study. *International Journal of Tourism Research*, 13: 570-582.
Gunn, C.A. (1994). *Tourism Planning Basics, Concepts. Cases*. Washington: Taylor & Francis.
Henderson, R. (1911). *Ninety years in the Master's Service: reminiscences of a pioneer evangelist*. Edinburgh: Andrew Elliot.
Hodgson, C.P. (1846). *Reminiscences of Australia: with hints on the squatter's life*. London: W.N. Wright.
Horne, J. (2005). *The Pursuit of Wonder: How Australia's Landscape was Explored, Nature Discovered and Tourism Unleashed*. Melbourne: The Miegunyah Press.
Howitt, R. (1845). *Impressions of Australia Felix* ... London: Longman, Brown, Green & Longmans.
Inglis, A.S. (2007). *Summer in the Hills: the Nineteenth-Century Mountain Resort in Australia*. Melbourne: Australian Scholarly Publishing.
Jacobsen, J.K.S. (1997). The making of an attraction: the case of North Cape. *Annals of Tourism Research*, 24(2): 341-356.
MacCannell, D. (1976). *The Tourist: A New Theory of the Leisure Class*. New York: Schoken.
Mossman, S. & Banister, T. (1853). *Australia, Visited and Revisited*. London: Addey & Co.
Ousby, I. (2002). *The Englishman's England: Taste, Travel and the Rise of Tourism*. London: Plimlico.
Ryan, S. (1996). *The Cartographic Eye – how explorers saw Australia*. Cambridge: Cambridge University Press.

Slade, P. (2003). Gallipoli thanatourism: the meaning of ANZAC. *Annals of Tourism* Research, 30(4): *779-794*.
Smiles, S. (ed.) (1880). *A Boy's Voyage Round the World* ... London: John Murray [First published 1871].
Towner, J. (1996). *An Historical Geography of Recreation and Tourism in the Western World 1540-1940*. Chichester: John Wiley & Sons.
Twopeny, R. (1883). *Town Life in Australia*. London: Elliot Stock.
Young B. (1983). Touristisation of traditional Maltese fishing-farming villages. *Tourism* Management, 3: 35–41.

2 Lal Lal Falls Scenic Reserve by Ian D. Clark

Lal Lal Falls near Ballarat in Western Victoria evolved over 162 years from an Aboriginal cultural site into a recreational and tourism attraction. Key moments in this history were the visit by two European Port Phillip Aboriginal Protectorate officials in 1840, the reservation of the site as a public park in 1865, and the tragic death of two school children from a landslip at the site in 1990. To understand the history of Lal Lal Falls visitation, this study uses perspectives developed by MacCannell (1976), Butler (1980), and Gunn (1994). MacCannell's (1976) research into the development of secular attractions through five stages – sight sacralization or naming, framing and elevation, enshrinement, duplication, and social reproduction – will be tested to see if it satisfactorily accounts for the development of the Lal Lal Falls attraction. Butler's (1980) tourism area life-cycle model may explain the subsequent stagnation and decline of the attraction, particularly since the 1990 catastrophe. Gunn's (1994) spatial model of attractions should be able to add a spatial dimension to understanding the history of recreation planning at Lal Lal Falls in terms of three zones (nucleus, inviolate belt, and zone of closure) of visitor interaction outlined by the model. This chapter extends earlier research by Clark (2002) into the history of the waterfall.

The Lal Lal Falls are one of two falls found on Lal Lal Creek, a tributary of the Moorabool River, and is situated in the Parish of Lal Lal, and in the Moorabool Shire, three-kilometres east of Lal Lal township, and twenty-five kilometres from Ballarat (see Fig.2.1). The second waterfall on Lal Lal Creek is found near the site of the Peerewur Pastoral homestead.[1] Lal Lal Creek joins the Western Moorabool River, approximately one km downstream of the falls, and travels another 300 metres before entering the Lal Lal Reservoir, which was established in 1972. The falls, formed when a volcanic basalt tunnel collapsed, cascade 34 metres to the pool at the base. The falls are situated on Pleistocene basaltic plains formed about 1.8 million years ago. The lava plain originated from Mount Buninyong (Land Conservation Council [LCC], 1980).

The Lal Lal Falls is situated within the traditional country of the Wathawurrung Aborigines, and is one of Victoria's most significant Aboriginal cultural sites, as it is one of several recorded living sites of Bunjil – the Kulin peoples' creator spirit (Bonwick, 1863; Massola, 1957, 1968, 1969).

The aesthetic appreciation of waterfalls and the spectacles they offer is a landscape taste that has existed in some cultures for over two thousand years (Hudson, 1998: 959). Waterfalls are not ubiquitous and Hudson (1998: 959) has suggested that

1 Osborn (1973:48) suggested the name of this waterfall was Peerewur, which meant 'waterfall' and 'possums'. An earlier name for the falls is 'The Mustangs', in reference to where the Peerewur station's mustangs were kept. The Peerewur Falls have a 3.6m fall, whereas the drop of the Lal Lal Falls is 34.16m.

as tourism resources, along with other natural curiosities such as caves, geysers, and glaciers, they have been neglected in the tourism literature.

2.1 First Phase: Sight Sacralization and Naming 1837–1846

In terms of MacCannell's (1976) first phase in the development of attractions, that of 'sight sacralization' or 'naming', three Aboriginal names have been recorded for the falls – Bunjil, the creator spirit; Woringganninyoke, meaning unknown; and Lal Lal, said to mean 'dashing of waters in a crevice' (Clark and Heydon, 2002). An early historian of Ballarat, William Withers (1999: 4), claimed that the name Lal Lal had the following origin:

> Before Mr Pettett took up the Dowling Forest run [in 1838] he was living at the Little River, and a native chief named Balliang offered to show him the country about Lal Lal. The chief in speaking of it distinguished between it and the Little River by describing the water as La-al La-al – the *a* long – and by gesture indicating the water-fall now so well known, the name signifying falling water.

Balliang was a clan-head of the Wathawurrung baluk, a Wathawurrung clan of the Barrabool Hills near Geelong, whereas the Lal Lal Falls formed part of the country of the Wathawurrung clan known as Tooloora baluk whose country encompasses Mt Warrenheip, Lal Lal Creek, and Meredith (see Clark, 1990). The name 'Lal Lal' appears to be a generic Wathawurrung word for waterfall, and is not the specific name of this feature, however; the word was sourced from another Wathawurrung speaker but wrongly conferred by Europeans.

Local historian Nathan Spielvogel claimed 'Lal Lal, correctly pronounced "La-al La-al" was the aboriginal name for "the big crack" referring of course to the Falls' (Mansfield, 1982:157); however, it is apparent that Spielvogel was unaware of the names recorded by Robinson in 1840 and 1846.

Europeans are thought to have first learned of the existence of the falls in 1837 when Frederick D'Arcy, a Government Surveyor, was working with his party on the upper reaches of the Leigh and Moorabool rivers, and was told of the falls by local Aborigines (Griffiths, 1988:1). D'Arcy was among a party of squatters in August 1837 that was formed with the purpose of exploring the Buninyong district. The party included Dr Alexander Thomson, George Russell, David Fisher, Captain Hutton, Thomas Learmonth, and Henry Anderson and an Wathawurrung Aboriginal guide possibly named 'Darriwill' (see Macqueen, 2010: 111f). They visited the Lal Lal Falls, only to find it was not running owing to the dryness of the season. George Russell, one of the participants, explained the context of the visit to the falls:

> [In early spring of 1837 one of the government surveyors Frederick d'Arcy, who was sent to the Geelong district to survey the principal rivers and creeks and note remarkable objects] received

> information from the natives about a fine waterfall at the upper part of the Moorabool River, which the natives called Lal Lal. A party was made up in order to visit this waterfall and explore a little of the country round it ... By arrangements we all met at Mr Anderson's hut one evening, near Russell's Bridge, where we stopped for the night; and we started for Mr d'Arcy's camp.
>
> We got on his dray-tracks, and after passing where Meredith is now situated arrived at the camp during the afternoon, and spent the night there. Mr d'Arcy having had an extra tent pitched for us. After an early breakfast we started for the falls, Mr d'Arcy having given his men orders to strike the tents, to put them and everything else on the dray, and proceed to Mount Buninyong, which was in sight of the camp, there to wait at the foot of the mount until we joined them.
>
> After leaving camp we kept pretty close to the Moorabool river, following it up and passing through some rather thickly wooded and rangy country, reaching the falls about midday. But there was no waterfall. Owing to the dryness of the season the water had not commenced to run in the Moorabool, although it was late as the month of August. All that we saw was a steep precipitous rock, about eighty feet in height. I have since heard that the waterfall is worth seeing when the river is full (Brown, 1935: 158).

Withers claimed that 'Before Mr Pettett took up the Dowling Forest run he was living at the Little River, and a native chief named Balliang offered to show him the country about Lal Lal. The chief in speaking of it distinguished between it and the Little River by describing the water as La-al La-al – the a long – and by gesturing indicating the water-fall now so well known, the name signifying falling water' (Withers, 1870: 4). William Pettett was superintending at 'Dowling Forest' at Lake Learmonth for WJT Clarke from 1838. The D'Arcy version is sourced from one of the participants (George Russell); the source for Withers' account is unknown.

This discovery and visit in August 1837 by at least seven Europeans suggests that these seven may potentially serve as the agency of the spread of news about the waterfall's existence. The fact that it was not running when they visited, but they considered it worth seeing when the water was flowing, may have led to revisits by some of the individuals of this party, though there is no evidence for this. Qualifying the spread of information, however, is the local Ballarat newspaper editorial of 1857 (see below) that reports that the falls were not well known; so there appears to have been minimal word of mouth recommendation of its existence. Once the falls were incorporated into a pastoral lease, visitation appears to have been mediated by the lessee. The Lal Lal Falls formed part of Hugh Blakeney's and Charles Ayrey's Lal Lal Pastoral Run that they took up in early 1840. In 1843, Peter Inglis purchased the Lal Lal lease. According to Billis and Kenyon (1974), John Whitehall Stevens was lease holder from 1845-6; though there are no references to this in local histories. From 1846, Archibald Fisken, Inglis' nephew, was given the management of Lal Lal, and eventually took ownership in 1854.

The falls were visited on 8 March 1840, by the Chief Protector, George Augustus Robinson, and Assistant Protector, Edward Stone Parker, public officers of the department of the Port Phillip Aboriginal Protectorate. The falls fell within Parker's jurisdiction, the Loddon or Northwestern District. Robinson and Parker had spent the previous night at Charles Ayrey's and Hugh Blakeney's Lal Lal station. Robinson, in his journal, advised that Ayrey had told him about the falls during the evening and that he guided them to the waterfall. This confirms that Ayrey is the agency through which the falls became the focus of Robinson's and Parker's visitation.

Robinson's account of his visit to the falls provides some confirmation that the name Lal Lal is not a local name, as implied by Withers. Robinson recorded Aboriginal names for Lal Lal Creek and names for both of the falls found on the creek. Robinson's account of his visit is as follows:

> Sunday 8 March 1840: A.M. Went with Mr Airey to see the fall of the western branch of the Marrabul, which Mr Airey informed me of the evening previous, about half of a mile from Mr Airey's residence. This is the finest fall I have yet seen in the country. It falls a depth of 80 feet. The rock is basalt, the best formation I have seen in this province being pentagonal and distinctly marked. The sides of the ravine, as well as the head which forms the fall, are perpendicular; the S. side is concave, whilst the N. is short. The fall is quite abrupt. The channel along which the water from the surrounding country is brought to the fall is shallow, a mere hollow in the surface from which in wet season a considerable body of water discharges itself to the depth of 80 feet over the precipitous cliff which forms the head of the ravine. And then over a rapid a few hundred yards lower down. This fall in wet seasons must be magnificent (Clark, 2000a:199).

Robinson had recorded Woringganninyoke as the name of the Lal Lal Falls, and Nanden as the name of the little fall, presumably a reference to the much smaller fall upstream. The name for Lal Lal Creek was recorded as Yarkmyowing by Robinson. These names would seem to confirm that Lal Lal is a generic Wathawurrung name for waterfall.

Although Robinson in 1840 does not discuss the cultural associations of the site, Parker alluded to the existence of an Aboriginal legend in an article he published in the *Port Phillip Herald* after the March visit. *The Port Phillip Herald* article was reprinted in the *Morning Chronicle* (Sydney) (8 February 1845), *The Australian* (6/2/1845), *Launceston Advertiser* (14/2/1845); and *The Sydney Morning Herald* (3/10/1846). This article was republished in Braim's (1846) history of New South Wales. Parker wrote, 'Other traditions exist among them referring to the origin of certain natural objects. Thus they believe in the existence of another mythological being called Bonjil or Pundyil, who, however, is said to have been once a "black fellow", and a remarkable locality is indicated as his residence when on earth. This is the deep and basaltic glen or hollow, forming the fall of "Lallal" on the Marrabool, near Mr Airey's Station. He is now represented as dwelling in the sky, and it is curious that they call the planet Jupiter "Pundyil", and say it is the light of his fire. This Pundyil is said to have found a single kangaroo, emu, and other animals on earth: that he caught them, cut them up, and by some mysterious power, made each piece into a new kangaroo, &c., and that hence the country was filled with these animals (Braim, 1846 vol. 2: 444-445).

According to Billis and Kenyon's (1974) register of pastoral holdings, Robinson and Parker confused Lieutenant John Moore Cole Airey, at 'Happy Valley' in 1839-53, with Charles Ayrey who was with Hugh Blakeney at 'Lal Lal' during 1840-41; however, Withers (1870: 5) and Griffiths (1988: 5) identified Blakeney with George Airey, the brother of Lieut. Airey.

Robinson confirmed the falls connection with Bunjil when he visited the site a second time, on 7 August 1846, and this time the creek was flowing and water was falling. On this occasion Robinson did not stay at the Lal Lal station and he went to the falls without any guidance from local squatters, however, he was accompanied by a Djadjawurrung man named Merrigundidj who joined Robinson at the Mt Franklin protectorate station on 11 April 1846, and travelled with him from then until 8 August 1846, when he returned to the Loddon station. Robinson, Merrigundidj, and a border policeman named Patrick Farrell, went down the Murray River to Tyntynder station at Swan Hill and from there into South Australia where they visited the copper mines at Kapunda, the Native School in Adelaide, and returned to Port Phillip via Encounter Bay and Rivoli Bay and the Mount Rouse Protectorate station:

> Visited the falls of Lal Lal, one which forms the gap large and the other part falls a height of [blank] width of [blank], was in all its glory worth seeing basaltic column rapid down another glen runs to it. ... Natives call it Punjil, so Merryonedeet says, said [blank] belonged to it, was at Buninyong camp (Clark, 2000b:108).

Bunjil was also associated with a cave at Cape Schanck. Howitt (1845) and Bonwick (1863) both discuss this cave:

> On this beach is the most remarkable natural cavern yet discovered in Australia Felix. ... Altogether it is a wonderful cave. There is also as singular a tradition about it. As Pungil, the god of the aborigines, say the natives, was one day taking a walk on the sea, suddenly there came on a storm; when coming to the rocky shore, he spoke to it, and immediately, at his word, the rocks rose up, and this cave was fashioned before him. Into it the god stepped, and sheltered until the tempest was over (Howitt, 1845: 147-8).

James Bonwick (1863: 54), in discussing the Cape Schanck cave, also discussed Bunjil's residence at Lal Lal Falls:

> At Cape Schanck, of Western Port, a cave is pointed out from which Pundyil or Bin-Beal used to take his walks beside the sea. He was accustomed when upon earth to frequent other caves, chasms, or dark places. Deep basaltic glens were favourite homes. We are well acquainted with one of these assumed divine residences situated in a romantic volcanic rent some fifteen miles from Ballaarat, through which the river passes after rolling down the Lal Lal falls. The planet Jupiter shines by the light of his camp fire in the heavens, whither he has now retired.

Parker's 1840 publication of the Aboriginal cultural associations surrounding Lal Lal Falls, and its republication in 1845 and 1846 should have ensured that information about the falls and its significance would be distributed wider than local squatters and their employees, and yet a survey of accounts of travellers to the district reveals that this does not appear to have occurred. Three accounts from travellers who visited the Ballarat district in the 1850s and 1860s (Clacy, 1853; D'Ewes, 1857; Snell, 1988) discuss the falls and only two directly visited them (D'Ewes, 1857; Snell, 1988). The Snell journal reference remained unpublished until 1988.

Ellen Clacy (1853: 133), learned of the existence of the falls during an 1852 visit to the Ballarat goldfields, but failed to visit them.

> The country round Ballarat is more in the North American style, and when the creek is full, it is a fine sight greatly resembling, I have heard, one of the smaller rivers in Canada; in fact, the scenery round Ballarat is said to approach more to Upper Canada than any in the colony. The rocks, although not high, are in places very bold and romantic, and in the wet season there are several water-falls in the neighbourhood (Clacy, 1853: 133).

John D'Ewes visited the Lal Lal Falls in August 1853 during a three-month *locum tenens* appointment as Police Magistrate at Ballarat. During this time he took the opportunity to make excursions into the surrounding countryside 'in search of picturesque scenery or sports, or the two combined' (D'Ewes, 1857: 49). With reference to the falls, he wrote:

> On the banks of a stream, was situated the pretty station of Lallal, belonging to a Mr. F....N, a thriving and hospitable squatter; near this station was a celebrated waterfall, which, in point of picturesque beauty, for its size, surpassed any I ever met with. The width of the first fall was not above 100 feet, with a sheer perpendicular descent of the same extent. The granite formation on each side was exquisitely beautiful, and so varied and delicate in its shapes and tracery that it might have passed for the handywork of some skilful artificer, as indeed it was, and the greatest of all – Nature! From this fall a succession of similar cascades appeared in a long vista, with an intervening space of several hundred yards between each, as far as the eye could reach. The deep clear basin at the bottom of each fall was full of a peculiar description of eel, some of enormous size, and weighing as much as ten pounds. On my first visit to this place I managed to scramble down the rocks with some rough fishing tackle, for the purpose of procuring a dish of the eels in question, and adventure attended with some danger on account of the snakes that infest this country (D'Ewes, 1857: 56-7).

In this entry D'Ewes is referring to Archibald Fisken, manager at Lal Lal for his uncle Peter Inglis, 1847-50; then licensee from 1850 onwards. The Lal Lal run, 18,313 acres, was first taken up by JW Stevens in 1845; Inglis took over the lease in October 1846.

Engineer Edward Snell visited Lal Lal Falls on 12 November 1855, and recorded the following account in his journal:

> We had some luncheon at the Corduroy Bridge Hotel (the last 5 miles by the bye was over a plank road,) and then we struck off into the bush, steering North East for Lal Lal falls. We made the Lal Lal Creek 2 miles above the falls owing to our not getting the exact compass bearing and then

> tracked the Creek down until we came to the falls. I made a sketch of them from the top and afterwards descended into the ravine where we sat down, had dinner, grog, and a smoke and I sketched the view on the opposite page. We then visited the Little Falls on a creek about a mile further on and after a smoke and more grog started for Egerton ... (Snell, 1988: 360).

Snell arrived in Australia in 1849, and it is not possible to determine how he knew of the falls, perhaps he was told about them at the Corduroy Bridge Hotel, at what is now Clarendon, or he knew of them through Parker's or Braim's publications, nevertheless his directions were somewhat vague and he and his party overshot them and had to travel down the creek to locate them.

Anthropologist, Aldo Massola (1968: 59), described the Kulin 'myth' of the 'Lal-Lal Falls on the Moorabool River' thus: 'Bunjil made the falls to relieve the monotony of the landscape. He liked them so much that he decided to make them his earthly home'. This story is unsourced, however Massola (1968: x) explained in his foreword that the accounts he published 'were collected over a period of ten years, from Aborigines in all parts of Victoria', and he supplemented these with 'the scant published material'. Other sites directly connected with Bunjil include 'caves' at Bushy Creek and Cape Schanck, and the rock-art site known as Bunjil's Shelter in the Black Range near Stawell (Massola, 1957). Massola speculated that Bunjil chose to live at the falls because of its idiosyncratic features:

> Apart from the fascination of watching the wide creek ending its placid run through the level plain by suddenly tumbling, with a mighty roar, down the 200 feet chasm, there were other reasons, no doubt, why Bunjil was made to live there. One was the fact that the swamp supported a large population of birds and other animals which assured the Aboriginals of plentiful supplies of food. Another was the comfort of the sand dunes on the south-east of the swamp, which make ideal camping places. A third, and no doubt very important reason, was the deposits of white pipe-clay on the east side of the swamp, which are now commercially quarried for paper clay (Massola, 1969: 70-71).

Charles Barrett (1944: 110) argued that Victoria is lacking in 'majestic waterfalls', though it does possess 'some beautiful falls', some of which have impressed international visitors.[2] Yet Robinson, who visited the site in 1840 and 1846 considered the Lal Lal Falls were 'magnificent', and 'the finest fall' he had 'seen in the country'. He also found the geology of the site 'the best formation' he had seen. Its significance lies in its cultural and natural values, yet apart from providing a name for the falls, Aboriginal cultural values were not instrumental in the sacralization of the site. Natural values provided the greatest interest in that process.

The aesthetic appreciation of waterfalls and the spectacles they offer is a landscape taste that has been documented to have existed in some cultures for over two thousand years (Hudson, 1998: 959). Hudson (1998: 959) has noted that waterfalls as

2 Rudyard Kipling, for example, was impressed with Erskine Falls and wrote a poem about Lorne.

recreation and tourism resources 'have been largely neglected in scholarly literature', which is surprising given that they have scarcity value. 'Unlike fire, clouds and wind-tossed trees, however, waterfalls are not ubiquitous and, indeed, may be regarded as curiosities of nature along with caves, geysers, and glaciers' (Hudson, 1998: 963).

2.2 Second Phase: Framing and Elevation 1865–1885

The second phase identified by MacCannell (1976) in the evolution of attractions is 'framing and elevation' which he argued results from an increase in visitation, when demand requires some form of management intervention, whereby the sight is displayed more prominently and framed off. In the case of Lal Lal Falls this phase began in 1865 and ended with the construction of the Lal Lal to Lal Lal Racecourse railway branch line in 1885. Reservation of the site in 1865 ensured that the falls would not be damaged by incompatible land use or allow them to encroach upon the nucleus and compromise the visual entrance to the falls.

The Lal Lal Falls formed part of Blakeney's and Airey's Lal Lal Pastoral Run which they first took up in early 1840. In 1843, Peter Inglis purchased the Lal Lal lease. From 1846, Archibald Fisken, Inglis' nephew, was given the management of Lal Lal, and eventually took ownership in 1854. In 1857, freehold land became available as the government brought it up for auction. Section 8 of the surveyed section contained both the Moorabool and Lal Lal falls. *The Star*, a local Ballarat paper, in an editorial dated 19 September 1857 brought the sale to its readers' attention, but the opening line reveals that local awareness of the falls was not widespread.

> Probably few of our readers are aware that within ten or twelve miles of the township of Ballarat, there are two fine waterfalls, one of which from its picturesque beauty and its great height would not be unworthy of attention even in the best parts of the Scottish Highlands. These falls, the Lal Lal and the Moorabool, are situated in a beautifully undulating and finely timbered country on the south-eastern base of Warrenheep, from and around which mount, the streams which from them take their rise.

The editor did not begrudge Archibald Fisken receiving a fair compensation for the various improvements he had made to the section being sold. The *Miner and Weekly Star* (25/9/1857) made the following comment on the proposed sale of the Lal Lal Falls:

> But, however, the question of how the area of the lot (537 acres) may be ultimately disposed, the Government should at once withdraw it from the present sale for the purpose of resurveying them, and of preserving to the public the beautiful waterfalls we have alluded to, as well as the extensive water frontages that by the present plan would in the event of a sale be for ever alienated.

The following week *The Star* (26/9/1857) increased the pressure on the Victorian Government to set aside the Lal Lal Falls when it published an account of a visit to the waterfall:

> A traveller sends us the following description of these beautiful falls: On coming forward to a view of the Lal Lal Falls, which from the flatness of the country is done all at once, the eye suddenly beholds a cascade of water rushing over a precipice of rock 110 feet perpendicular height, into a ravine, lined on both sides near the top with basaltic columns, identical with what are to be seen at the Giant's Causeway in Ireland, and at Staffa and Iona in Scotland; in every way shaped the same, being irregular polygons with from three to eight sides, set into each other with convex and concave joints and all standing perpendicular. From having seen these natural curiosities in the old country, which thousands visit annually from all parts of the world, it may be easily supposed that never having read of, or heard described the scenery of Lal Lal, that when I first saw these Pillars, I was struck with astonishment and wonder at the similarity; particularly when it is considered that the places are distant from each other 14,000 miles, I felt as if I had suddenly been brought in contact with old and familiar faces. But to return to your editorial remarks of 19th inst., I would beg leave to suggest, that when the Government do sell the adjacent lands, that they bind the purchasers to leave open roads through them to the Falls, so as the lovers of natural scenery may be able to ride, drive, or walk from Ballarat, Buninyong, Corduroy, and Mount Egerton (from which places the Falls are about equi-distant), without let or hindrance from the fortunate possessors of that part which surrounds decidedly the finest scenery and the greatest natural curiosities in this part of the Colony of Victoria.

There is little doubt that the primary motivation in seeking the preservation of the Lal Lal Falls was its aesthetic value – the scenery of its landscape – and in this the moves to reserve them paralleled moves in North America to preserve wilderness in places such as Yosemite, Yellowstone, and Niagara Falls (Towner, 1996:160) (see Fig.2.2). *The Australian News for Home Readers* (25/7/1864) even went so far as to suggest the Lal Lal Falls were a miniature Niagara:

> These falls, which are situated in a wooded country near Ballaarat, form the prettiest water scene the locality can boast. The watercourse is suddenly broken by a precipitate fall of the earth, almost as perpendicular as a wall. The water leaps down a great number of feet, forming a miniature representation of Niagara. The falls are a favourite resort for picnic parties, the surrounding country presenting many picturesque points, and they are usually visited by strangers among the other "lions" of the Ballaarat district.

The Government subsequently withdrew the Lal Lal Falls section from public sale, and 200 acres (90 hectare) were permanently reserved as the 'Lal Lal Park' in 1865.[3] Norman (1984: 2) states the reserve was proclaimed in 1877 (Victorian Government Gazette, 1877: 1652). Its formal name is 'Lal Lal Falls Scenic Reserve'. The purpose of scenic reserves is to preserve scenic features and lookouts of particular significance (LCC, 1980). From 1877 until 1964, the Scenic Reserve was under the control of a com-

3 Norman (1984: 2) states the reserve was proclaimed in 1877 (Victorian Government Gazette, 1877: 1652).

mittee of management of four representatives from the Shire of Buninyong and the City of Ballaarat. From 1964, the Shire of Buninyong had sole management for the reserve (Victorian Government Gazette, 1964). From 1977, an officer with the West Moorabool Water Board was serving as unofficial caretaker of the reserve (Norman, 1984: 19).

An adjacent reserve known as the 'Racecourse Reserve', to the south of the Scenic Reserve, has been temporarily reserved as a possible car-park extension (Victorian Government Gazette, 1969: 1823). The Lal Lal railway station was constructed in 1862, and the *Ballarat Star* (26/4/1862) questioned its purpose, although it considered it 'may prove however to be useful to picnic parties during the summer months'. Annual races were first held at the Lal Lal Racecourse Reserve in 1864.

2.3 Third Phase: Enshrinement 1885–1938

MacCannell (1976) has identified 'enshrinement' as the third phase in the evolution of attractions. Critical in the enshrinement phase of Lal Lal Falls is the Lal Lal picnic races. This phase began when access to the racecourse was improved in 1885, when a branch railway line was opened from the Lal Lal station that went directly to the racecourse. A spacious grandstand was erected in 1887, capable of accommodating 1,000 patrons (Griffiths, 1988: 112). The number of visitors to the Lal Lal races evidences the rapid tourism growth associated with this phase. In 1911, for example, special trains took 1,303 individuals to the Lal Lal picnic races (Bate, 1993: 152). Griffiths (1988: 112) described the New Year's Day races at Lal Lal as an 'institution', with special trains catering for up to 30,000 visitors from Ballarat and district:

> Many of the crowd came just for a picnic day, and remained in the outside reserve to enjoy themselves at the many side shows and boxing tent. Inside the course were office, kiosk, publicans' booths, and plenty of bookmakers. The cover of the official programme warned punters to 'Beware of Welchers, Bet Only with Recognised Bookmakers'.

According to Griffiths, the Lal Lal Turf Club held its principal meeting on New Year's Day annually until 1936, after which it chose another day to avoid its clash with the Burrumbeet meeting. The last official meeting was 31 December 1938. Although several gymkhanas were held after 1938, the improvements were sold and the branch railway torn up and the racecourse reserve is now a paddock leased by the Shire Council.

2.4 Fourth Phase: Duplication

MacCannell's (1976) fourth phase in the development of attractions is that of 'duplication', when copies of the nucleus of the attraction, in this case the waterfall, are made available through media such as paintings, photographs, and postcards (see Figs. 2.3 & 2.4).

The falls have been associated with prominent artists and photographers, including the German romantic colonial artist Eugen von Guerard who drew the falls in 1858, the Richard Ledger photograph taken some time after 1870, and the 1977 oil painting by Fred Williams.[4] These are all examples of the influence of art on developing aesthetic tastes for landscape. Von Guerard's drawing 'Fall of the Lallal Creek' is based on scenes he had sketched during 1853 and 1854 when he was gold mining in the Ballarat district (Tipping, 1982: 10). Tipping (1982: 10) claimed that von Guerard was attracted to places such as Lal Lal Falls because they 'relieved the monotony of the basalt plains' and 'symbolised the violent and eroded forces of nature that had created them'.[5]

In the 1850s, the demand for black-and-white art was significant, and the Lal Lal Falls drawing is one of several landscapes that were reproduced by wood engraving. The *Illustrated Melbourne News* on 16 January 1858 reproduced the drawing from wood engraving. Von Guerard compiled a portfolio of sketches and newspaper cuttings, which included the Lal Lal Falls reproduction, and named it *Australien Reminisunzen*; it is now in the Mitchell Library in Sydney (Bruce, 1980). Von Guerard's aesthetic sensibility with its romantic and picturesque taste for scenes based around mountains, gorges, forests, and waterfalls, calls to mind Rousseau's declaration in his *The Reveries of a Solitary* of 1778, that 'No flat country, however beautiful it may be, ever appeared so to me. I need torrents, rocks, firs, dark woods, mountains, rough tracks to climb up and down, precipices by my side to give me a nice fright' (quoted by Towner, 1996:140).

In relation to waterfalls and advent of photography, Reilly and Carew (1983: 80) have noted that 'Waterfalls were a common subject for photography in the nineteenth century and views of the Niagara Falls, especially, were in great demand throughout the world. In Australia from the 1860s onwards there was hardly an album or folio of photographs that did not include a view of a waterfall'. This is confirmed by the Fauchery-Daintree collection of 1858 that includes photographs of Turpin Falls and Glen Lyon Falls.[6]

---

4 Other photographs may be found in the National Library of Australia – see www.nla.gov.au.

5 Other waterfalls sketched and painted by von Guerard include Strath Creek Falls, The Weatherboard Falls, Clyde Fall, Moorabool Falls, and Wannon Falls.

6 This collection is in the State Library of Victoria

2.5 Fifth Phase: Social Reproduction

For MacCannell, the final stage of sight sacralization is social reproduction which 'occurs when groups, cities, and regions begin to name themselves after famous attractions'. In the case of Lal Lal Falls, social reproduction has occurred in the naming of the Lal Lal pastoral run now known as Lal Lal Estate, Lal Lal township, Lal Lal reservoir, Lal Lal railway station, Lal Lal primary school, and businesses such as Lal Lal Wind Farm, and Lal Lal Demolitions in nearby Ballarat. There are only two businesses that have reproduced the full name of the attraction – Lal Lal Falls Hotel, and Lal Lal Falls Cottage. In terms of the other Aboriginal names of the falls, Bunjil survives in the nearby Bungal Dam and the Bungall pastoral run on Lal Lal Creek adjoining Lal Lal first taken up in April 1838 by George Egerton.

2.6 Tourism at Lal Lal Falls

In 2001, Beggs-Sunter (2001: 3) summed up the history of tourism at Lal Lal Falls in the following terms: 'One of the most significant heritage sites in our region is the Lal Lal Falls, once a favourite attraction, now virtually forgotten'. Thorpe and Akers (1995: 64-5) confirmed that, during the Christmas season, the Lal Lal Falls and Racecourse were popular family picnic areas. The Buninyong community subsequently planted conifers and constructed a rotunda along with barbecue facilities and toilet facilities. In 1969, Massola (1969: 71) described the falls thus:

> With the passing of the Aborigines, Lal-lal Falls have become the top tourist attraction of the Shire of Buninyong, and have been developed accordingly. On the high ground adjoining the falls a picnic and caravan park has been laid out, and concrete steps, complete with safety railing, providing access down into the valley and to the base of the falls, have been built.

The reserve also has significant natural values, particularly the northern side of the Lal Lal Creek, approximately 500 metres downstream from the falls. In 1980, Lal Lal Falls were listed as one of several sites in the Ballarat region where sign-posted walking tracks were popular with day travellers and family groups (LCC, 1980: 148). The falls were also highlighted as a popular feature for recreational driving: 'For many people, driving to an area is part of the recreational experience, while for others it simply provides access to areas where they will carry out some other activity. Pleasure drivers, the former category, look for diversity in scenery and natural history' (LCC, 1980: 150). The Lal Lal Falls with their basalt columns are of interest to persons with some knowledge of geological phenomena associated with recent volcanic activity. Table 1.1 outlines the major management actions that have occurred along Lal Lal Creek and at Lal Lal Falls.

Tab. 1.1: Management actions at Lal Lal Falls and Lal Lal Creek

Year	Management actions
1863	Ballarat Water Committee dammed Lal Lal Creek, and Lal Lal Waterworks Association construct a 19km water race from Lal Lal Creek, just above the falls
1864	Annual races held at the Lal Lal Racecourse Reserve
1865	Government reserves 200 acres (90ha) as the 'Lal Lal Park'
1873	Lal Lal Turf Club forms
1885	Branch railway line opened from Lal Lal station to the Lal Lal Racecourse
1887	Grandstand constructed capable of accommodating 1,000 patrons
1938	Final Race meeting at the Lal Lal Racecourse
1939	Construction of walking bridge on the pathway to the Lal Lal Falls
1944	Lal Lal Turf Club disbands
1951	Walking track to the falls constructed
1953	Clearance of old trees in the picnic ground
1957	Installation of playground equipment
1961	Construction of lookout
1963	Construction of cement stairway to the falls and safety rail above the falls
1980	Reconstruction of the toilets, fireplaces, and seats and tables, and car park
1988	Fencing constructed at the Reserve to protect rare population of Discaria pubescens
1990	Access to the base of the falls is barred after a land slip kills two school children
2004-2008	Lal Lal Falls Advisory Committee undertook major upgrade of the site: removal of pine trees, upgrading barbeque facilities, improving interpretive signage and marking new walking tracks
2008	Walking track from Lal Lal Falls to Moorabool Falls opened

In 1984, an environmental consultant, Norman (1984) suggested thirteen management recommendations, the most significant of which were:

Protection of the Australian anchor plant;

Prevention of abseiling from the walls of the ravine;

Closure of the pathway leading to the summit of the falls;

Repair or replacement of vandalised amenities;

Provision of firewood on a constant, year-round basis;

Removal of old and dangerous pines and their replacement with selected English trees;

Improvements to the pathway to the base of the falls, and the construction of an appropriate sign to direct visitors to the pathway;

> Poisoning of problem weeds;
>
> Construction of a nature trail with plaques at select points that describe points of interest, from the picnic area to the base of the falls.

In 1988, a botanist from La Trobe University, in association with the Shire of Buninyong and the Department of Conservation, Forests and Lands, ensured that fencing was constructed at the reserve to protect a rare population of *Discaria pubescens* (Hairy Anchor Plant) under the *Flora and Fauna Guarantee Act* (1988). These plants were once common along the watercourses of the basalt plains. Since 1989, collection and seed propagation from the Lal Lal population has seen hundreds of plants introduced at fifteen localities in Western Victoria, including Mooramong Nature Reserve, Creswick Creek, and Carisbrook Reservoir.

Access to the base of the falls is via a concrete stairway and a walking track constructed in 1951. Norman (1984: 15) described it as a 'roughly made dirt path', that was 'dangerous and not suitable to the whole spectrum of visitors. It is very uneven and covered in some areas by large obstacles in the form of tree roots, thereby restricting physically handicapped people and parents pushing prams'. There is a second path at the reserve, one that runs north from the picnic area to the summit of the falls (individuals near the summit can be seen in Fig.2.1). A fenced in car park was constructed in 1980 that allowed vehicular access to the picnic area. In 1980, the concrete seats and tables at the reserve were in a state of disrepair.

Public discussion of site management by the Shire of Buninyong took place in January 1984. Councillor D. Leather stated that 'the Reserve was a popular site, but quite honestly, the place is a bit grotty'; Councillor A. Harbour argued that council 'should scale down its activities at the Reserve because the only use it was getting was abusive'; Councillor Leather disagreed, stating that 'many people used the Reserve and it required only little maintenance and, if kept on top of instead of neglected over years, then facilities would not be destroyed' (*Ballarat Courier*, 21/1/1984). The Ballarat Region Conservation Strategy 1999-2004 acknowledged in its section on Koori Cultural History that 'Lal Lal Falls was the special place of the great creator Bunjil (Bungal)'. The recommendations of the strategy were the provision of information to the general public by way of public lectures, school visits, and Koori heritage trail tours, and also to utilise the expertise of the Cultural Officer at the Ballarat and District Aboriginal Co-operative. In late 2001, the Moorabool Shire Council instituted a public meeting to discuss the future of the Lal Lal Public Reserve. Plans were announced to prepare a conservation management plan for the reserve. From 2004-2008, the Lal Lal Falls Reserve Advisory Committee, appointed by the Moorabool Shire Council, undertook a series of major projects at the reserve, including the removal of pine trees, upgrading visitor facilities such as barbeques and interpretive installations, and marking out new walking tracks. In early 2012, the *Ballarat Courier* undertook a survey of the Ballarat region's favourite natural attraction. Lal Lal Falls was considered the region's best, securing 37.5 per cent of the vote; Lake Wendouree came in second with 32.3 per cent, and Mt. Buninyong, third, with 9.4 per cent (*Ballarat Courier*, 8/2/2012).

2.7 Critical Moments in the History of the Lal Lal Falls

The falls and tourism there has been dramatically affected by three things: The diversion of water for gold mining in the 1860s; the cessation of the Lal Lal Races in 1938, and the death of two school children at the site in 1990.

In 1863, the Ballarat Water Committee dammed Lal Lal Creek so as to direct water across the Moorabool catchment for the use of gold miners in Ballarat. The Lal Lal Waterworks Association created other dams and these too affected the volume of water flow thus diminishing the spectacle of the Lal Lal Falls. This association constructed a 19-kilometre water race from Lal Lal Creek, just above the falls, to supply water to the miners at Dolly's Creek, and later extended to the Morrison's and Tea-Tree Diggings. The water from Lal Lal Creek was reported as having allowed Chinese diggers to 'turn over with profit every portion of the ground about Dolly's Creek containing the most minute particles of gold' (Mining Surveyors' Reports, October 1863). Although the literature fails to discuss the effect these diversions had on the flow of the water in Lal Lal Creek, the reduction in the volume of water that cascaded over the falls must, presumably, have been significant.

Another critical moment in its history as a tourist attraction came in 1990 when two school pupils died in a landslip whilst visiting the base of the falls. The Victorian State Coroner at the inquest found that there had been insufficient assessment of the region undertaken by the school's staff in view of the potential danger of rockfalls. The Coroner found that the college concerned had 'contributed to the deaths by undertaking a potentially dangerous activity without ensuring that appropriately qualified experts assessed the area before the crime' (Abrams, 2002). Until the 1990 tragedy, it was possible to walk along a track to the base of the waterfall; however, access is now restricted and visitors are able to see the falls from a viewing platform. The barrier has not been particularly successful as the walking track is still prominent, which suggests visitors are ignoring the signs barring entry and walking to the base of the fall. Lal Lal Falls was also a site renowned for abseiling, but after the 1990 landslip, this activity was banned.

2.8 Conclusion

This historical analysis of the Lal Lal Falls, near Ballarat, in Western Victoria, traces its evolution from an Aboriginal cultural site into a recreational attraction. The Lal Lal Falls, situated within the traditional country of the Wathawurrung Aboriginal people, is one of Victoria's most significant Aboriginal cultural sites, as it is one of several recorded living sites of Bunjil – the Kulin peoples' creator spirit. The Lal Lal Falls became a tourist attraction more for its natural significance than its Indigenous cultural values. Its reservation as a public reserve in 1865 marked the emergence of the site as an attraction with a nucleus and an essential setting.

Tourism at Lal Lal Falls experienced its zenith when it was associated with the annual Lal Lal Races and, with their cessation in 1938, the Lal Lal Falls Scenic Reserve, in terms of visitor numbers, has experienced continued stagnation and general decline; though this has now been arrested by major intervention works in 2004-2008. Since 1938, there have periods of management activity when site works have been undertaken, particularly in 1963, 1980, and 2004-2008; however, the benefits of the first two actions appear to have been short-lived and the infrastructure associated with the 'inviolate belt' – visitor amenities such as toilets, fireplaces, seats and tables – were vandalised and allowed to fall into disrepair.

What is particularly poignant about the Lal Lal Falls is its sacredness in Indigenous value systems – one of a select number of living sites of the creator spirit, Bunjil, before departing into the sky. Its significance cannot be overstated. Yet, although the Aboriginal significance of the site has been understated in previous site promotion and off-site interpretation,[7] visitor information and travel guides on the Internet are beginning to focus on the site's Indigenous values. For example, Walkabout, has the following entry for Lal Lal Falls listed under 'Things to see' in the Buninyong district: 'The local Kooris are said to have believed that Bunjil, their creator, resided at this place. The name is thought to be Aboriginal for 'dashing of waters'.[8]

In many respects the landslip in 1990 and the two resultant fatalities and the subsequent management decision to attempt to restrict visits to the base of the falls represents the lowest point in the tourism history of the attraction.

Lal Lal Falls, as a tourism attraction – one with significant natural and cultural values – has undergone a fundamental transition from being 'the top tourist attraction' and 'one of the most significant heritage sites' in the Buninyong region, where thousands of people would congregate for picnic races, to an attraction 'virtually forgotten' in 2001, only to reemerge in 2012 to be the leading natural attraction in the Ballarat district. This study has gone some way to explaining this transition, by showing the value of the models provided by MacCannell, Butler, and Gunn, and demonstrating that, when combined, they possess considerable explanatory potential for understanding the evolution and history of a tourism attraction. Tourism at Lal Lal Falls experienced its zenith when it was associated with the annual Lal Lal Races and,

---

7 Sources that discuss the site's indigenous values, such as Parker (1840), Robinson (1845 in Clark, 2000a), Braim (1846), Bonwick (1863), and Massola (1969), are not the kinds of publications that ordinarily inform populist tourism publications.

8 see Walkabout Australian Travel Guide [On-line:] http://www.walkabout.com.au/locations/VICBuninyong.shtml (accessed 4 December 2002)

with their cessation, the Falls, in terms of visitor numbers, has experienced continued stagnation and general decline until major site works in 2004-2008 which have seen the falls re-emerge in 2012 as the Ballarat region's most favourite natural attraction.

References

Abrams, N. (2002). Principal and Teacher Duty of Care for Students. Student Wellbeing Conference 14 June 2002:1-45 [On-line:] www.emr.vic.edu.au/disabwel/PDFs/Norman%20Abrams%20 legal%20Liability%20Issues%20in%20 School.PDF] (accessed 4 November 2002)

The Australian News for Home Readers 25/2/1864

Ballarat, (1999), Ballarat Region Conservation Strategy 1999-2004: a strategy for Sustainable Living [On-line:] http://www.ballarat.vic.gov.au/8heritage2.html (accessed 4 November 2002)

Ballarat Courier 21/1/1984; 8/2/2012

Ballarat Star 26/4/1862, 4/1/1864

Barrett, C. (1944). *Australian Caves, Cliffs, and Waterfalls*. Melbourne: Georgian House.

Bate, W. (1993). *Life After Gold Twentieth-Century Ballarat*. Melbourne: Melbourne University Press.

Beggs-Sunter, A. (2001). Lal Lal Falls in Newsletter, Buninyong and District Historical Society Inc., October, p. 3

Billis, R.V. & Kenyon, A.S. (1974). *Pastoral Pioneers of Port Phillip*. Melbourne: Stockland Press.

Bonwick, J. (1863). *The Wild White Man and the Blacks of Victoria*. Melbourne: Fergusson & Moore.

Braim, T.H. (1846). *A History of New South Wales from its Settlement to the Close of the Year 1844* (two volumes). London: Bentley.

Brown, P.L. (ed) (1935). *The Narrative of George Russell of Golf Hill with Russellania and Selected Papers*. London: Oxford University Press.

Bruce, C. (editor) (1980). *Eugen von Guerard*. Canberra: Australian Gallery Directors Council, Visual Arts Board of the Australia Council, Australian National Gallery, and Frank McDonald.

Butler, R.W. (1980). The Concept of a Tourist Area Cycle of Evolution: Implications for Management of Resources. *Canadian Geographer*, 24:5-12

Clacy, E. (1853). *A Lady's Visit to the Gold Diggings of Australia in 1852-53*. London: Hurst & Blackett.

Clark, I.D. (1990). *Aboriginal Languages and Clans – an historical atlas of western and central Victoria, 1800-1900*. Clayton: Monash Publications in Geography, Monash University

Clark, I.D. (ed.) (2000a). *The Journals of George Augustus Robinson, Chief Protector, Port Phillip Aboriginal Protectorate, Volume One: 1 January 1839-30 September 1840*. Clarendon: Heritage Matters.

Clark, I.D. (ed.) (2000b). *The Journals of George Augustus Robinson, Chief Protector, Port Phillip Aboriginal Protectorate, Volume Five: 25 October 1845-9 June 1849*. Clarendon: Heritage Matters.

Clark, I.D., (2002). The ebb and flow of tourism at Lal Lal Falls, Victoria: a tourism history of a sacred Aboriginal site. *Australian Aboriginal Studies*, 2:45-53.

Clark, I.D. & Heydon, T.G. (2002). *Database of Aboriginal Placenames of Victoria*. Melbourne: Victorian Aboriginal Corporation for Languages.

D'Ewes, J. (1857). *China, Australia, and the Pacific Islands* ... London: Richard Bentley.

Griffiths, P.M. (1988). *Three Times Blest – A History of Buninyong and District 1837-1901*. Buninyong: Buninyong and District Historical Society.

Gunn, C.A. (1994). *Tourism Planning Basics, Concepts. Cases*. Washington: Taylor and Francis.

Howitt, R. (1845). *Impressions of Australia Felix* ... London: Longman, Brown, Green & Longmans.

Hudson, B.J. (1998). Waterfalls Resources for Tourism. *Annals of Tourism Research*, 25: 958-973.

Land Conservation Council Victoria. (1980). *Report on the Ballarat Area*. Melbourne: Land Conservation Council Victoria.
Launceston Advertiser 14/2/1845.
MacCannell, D. (1976). *The Tourist: A New Theory of the Leisure Class*. New York: Schoken.
Macqueen, A. (2010). *Frederick Robert D'Arcy: colonial surveyor, explorer and artist c. 1809-1875*. Wentworth Falls, New South Wales: The Author.
Mansfield, P.G. (editor) (1982). *Spielvogel Papers Volume 1*. Ballarat: Ballarat Historical Society.
Massola, A. (1957). Bunjil's Cave Found. *The Victorian Naturalist*, 74:19-22
Massola, A. (1968). *Bunjil's Cave – Myths, Legends and Superstitions of the Aborigines of South-East Australia*. Melbourne: Lansdowne Press.
Massola, A. (1969). *Journey to Aboriginal Victoria*. Adelaide: Rigby.
Miner and Weekly Star 25/9/1857
Morning Chronicle 8 /2/1845
Norman, C. (1984). Draft Management Plan for Lal Lal Falls Scenic Reserve. Ballarat: Department of Applied Biology and Environmental Sciences, Ballarat College of Advanced Education.
Osborn, B. (1973). *The Bacchus Story – A History of Captain W. H. Bacchus, of Bacchus Marsh, and His Son*. Bacchus Marsh: Bacchus Marsh and District Historical Society.
Parker, E.S. (1840) article in *Port Phillip Gazette*
Reilly, D. & Carew, J. (1983). *Sun Pictures of Victoria The Fauchery-Daintree Collection*, 1858. Melbourne: Library Council of Victoria, Currey O'Neil Ross Pty Ltd.
Smiles, S. (ed.) (1880). *A Boy's Voyage Round the World* ... London: John Murray [First published 1871].
Snell, E. (1988). *The Life and Adventures of Edward Snell*.... North Sydney: Angus & Robertson.
The Australian 6/2/1845.
The Star 19/9/1857; 26/9/1857.
The Sydney Morning Herald 3/10/1846.
Thorpe, M.W. & Akers, M. (1995). *An Illustrated History of Buninyong*. Buninyong: Buninyong and District Historical Society.
Tipping, M. (ed.) (1982). *An Artist on the Goldfields The Diary of Eugene von Guerard*. Melbourne: Currey O'Neil.
Towner, J. (1996). *An Historical Geography of Recreation and Tourism in the Western World 1540-1940*. Chichester: John Wiley.
Walkabout Australian Travel Guide [On-line:] http://www.walkabout.com.au/locations/VICBuninyong.shtml (accessed 8 December 2002)
Withers, W.B. (1999). *History of Ballarat and Some Ballarat Reminiscences*. Ballarat: Ballarat Heritage Services [First published 1870].

Fig. 2.1: 'The Lal-Lal Falls Near Ballarat, Australia'. Source: The Graphic, 12 April 1873: p. 348.

Fig. 2.2: The Lal Lal Falls near Ballarat (1863); F. Cogne tinted lithograph. State Library of Victoria Pictures Collection H97.89/9

Fig. 2.3: Lal Lal Falls postcard (Ian Clark personal collection)

Fig. 2.4: Lal Lal Falls postcard (Ian Clark personal collection)

3 The Buchan Caves Reserve by Ian D. Clark

The Buchan Caves Reserve is some 360km east of Melbourne, near the township of Buchan. The Reserve is jointly managed by Parks Victoria and the Gunaikurnai Land and Water Aboriginal Corporation. It contains a visitor centre and facilities for overnight camping and day visitors. Within the reserve lies a honeycomb of caves with limestone formations – but there are only two show caves, Royal Cave (see Fig.3.1) and Fairy Cave (see Fig.3.2), and guided tours are conducted year round. The Buchan Caves Reserve falls within the Krauatungalung language area (Clark, 1998a: 189-190). This language or dialect, is one of five normally referred to as the 'Ganai nation' or 'Kurnai nation', a cluster of dialects sharing linguistic, social, cultural, political, and family associations.

3.1 First Phase: Sight Sacralization and Naming 1838–1907

In terms of MacCannell's (1976) first phase in the development of attractions as 'sight sacralization' or 'naming', the name Buchan Caves is derived from the place name Buchan which is found in the pastoral run name, the creek name, and the township name. The show caves within the Buchan Caves Reserve complex have their own names, which will be discussed in the next section. Within the ethno-historical records and general literature on Buchan, several meanings and derivations are given for the name 'Buchan'. George Augustus Robinson, the Chief Protector for Aborigines, for example, spelt Buchan several ways: Buckun (Jnl 3/6/1844 in Clark, 1998b); Buckin (Jnl 17/6/1844 in Clark, 1998b); Bucking (Jnl 21/6/1844 in Clark, 1998b); Buckan (Jnl 4/7/1844 in Clark, 1998b); and Bucken (Jnl 5/7/1844 in Clark, 1998b). There are several explanations of the derivation of Buchan, one that it is of Aboriginal origin, and another that it is Scottish. It is possible that it is polysemous, that is, that both have relevance.

One view is that it is of Scottish, or pseudo-Scottish, origin in light of the many people of Scottish origin who settled in the district and conferred the spelling Buchan after the town in Scotland (Howitt, 1904: 80; Seddon, 1994:63; Morgan, 1997: 21; Buchan Sesquicentenary Committee (BSC), 1989: 8). Another is that it derives from the Aboriginal word 'bukin' or 'bugin', a medicine-man of supernatural ability, dreaded because he stole human kidney fat its magical properties' (Roberts, 1977: 14); and that the caves were the haunt of the Bukin (BSC, 1989: 8). However, according to Howitt, the Baukan was an evil spirit of which little could be learned. He was only able to state they were negative, but not very powerful, and consequently not much feared (Fison & Howitt, 1880: 254). A third is that Buchan is a contraction of the Aboriginal placename Bukkan-munji (Howitt, 1904: 80); Bukan Munjie (Fison & Howitt, 1880: 192; Salierno, 1987:51), Bukkanmungie (Gardner, 1992: 17), Buk Kan Munjie, Bukinmunjie (Seddon, 1994:62). Howitt (1904: 80) noted that Bukkan-munji was 'the native name for the bag in which the Kurnai carries various articles', and literally means 'bag there' or 'the place of the bag'. According to Seddon (1994: 62) the name signifies a carrying bag, the common suffix 'munjie' indicating 'women's article'. Gardner (1992: 17) translates Bukkanmungie as 'place of the woman's bag'. The BSC (1989: 8), however, translate mungie as 'water'. Another translation is 'Grass bag' (BSC, 1989: 8).

William Thomas informed the Central Board for the Protection of Aborigines, in 1861, that 'Buccan' meant 'stack of rocks with a hole in it' (Pepper & De Araugo, 1985: 120). The Ganai word for 'cave' was most likely to be the equivalent of their word for hole, 'Ngrung' (Fison & Howitt, 1880: 191), which is used in reference to the 'hole of Nargun' (see chapter on Den of Nargun).

3.1.1 Indigenous Values at Buchan Caves

In 1952, Robert H. Lavelle, a writer born at Bairnsdale, submitted a draft manuscript entitled 'Buchan Caves Victoria Australia: Australia's most wonderful Caves' to the Department of Crown Lands and Survey.[9] The manuscript is of interest because it contains an interpretation of the significance of the caves, as well as a description of the site by an Aboriginal man named Harry Belmont.

> The Aboriginal tribes who inhabited these domains know of the caves from time immemorial, and for the reason that parts were used by the Headmen as a secret ritual ground, their existence was closely kept secret, and only by the chance of an inquisitive boy were they discovered in 1891 (Lavelle, 1952: 1).

Regarding the scenic setting of the caves, Lavelle (1952: 4) noted:

> The whole is a dream of Paradise, without saints; that is unless you are willing to admit that the natural fauna are the spirits of the Saints, as the aboriginals do. ... The best description I ever heard was uttered to me by an aboriginal friend, 'Harry Belmont', who pointed to the caves and said in his own tongue; 'Bogong, Murryang, Biamee'. I asked him for the translation of those three phonetic words. It is: Bogong: the birthplace of a great spirit. Murryang: we meet in that dream land at the end of the ocean where Biamee lives. Biamee: God, or Great Spirit. This, truly is the finest description one can give of the Buchan Caves: 'Paradise on earth'.

It has not been possible to learn anything about Harry Belmont. Preliminary analysis of the three Aboriginal words listed by Lavelle, as derived from Belmont, suggests a connection with New South Wales. Biamee is a reference to 'baayama' meaning 'god',

9 A copy is on file at the Parks Victoria Buchan Office. It was Lavelle's intention to submit the manuscript to the London magazine, *Wide World Magazine*. It seems it was never published.

listed in Austin's (1992: 53) dictionary of the Gamilaraay language of northern New South Wales spoken at places such as Moree, Narrabri, Gunnedah, Tamworth, and Boggabilla. Ridley (in Smyth 1878, Vol. 2: 285) confirms that Baiame (pronounced by-a-me) was used by Aboriginal groups scattered across northwest and west New South Wales, and was used by the Wiradjuri people at Mudgee and other localities. 'Bogong' is a Ngarigu word for the brown moth *Agrotis infusa* which breeds on plains in southern Australia (Dixon, Ransom & Thomas, 1992). Adult moths migrate to mountains where they collect in rock crevices in early summer and were harvested by Aboriginal people and were a staple food source at this time. This understanding of bogong diverges from that given by Belmont. It has not been possible to find any reference to Murryang.

When the Bataluk Cultural Trail was being developed in Gippsland in the mid-1990s (see below), the trail brochure stated the following about the Aboriginal heritage of the Caves:

> Traditionally Koorie people did not venture deep into the limestone caves at Buchan. There were, however, many stories about the wicked and mischievous Nyols which live in the caves below the earth.

Phillip Pepper, an Aboriginal elder at Lake Tyers, confirmed an Aboriginal involvement with the 'discovery' of one of the Buchan Caves. He recalled that in the early 1900s, his father, Percy Pepper was a friend of Frank Moon. They shared a passion for foot running, and would often run together. They also went rabbiting together. Phillip recalled that his father was setting rabbit traps with Frank Moon on one occasion when Moon 'found one of the caves at Buchan' (Pepper & De Araugo, 1989: 53).

In Gippsland, caves are associated with two mythical beings; the Nargun and the Nyol. For more information on the Nargun see the chapter on the Den of Nargun. Massola (1968: 74-5) has recounted the story of a Nyol at Murrindal.

> Once, when the tribe was camped at Murrindal, one of the men went possum hunting. Possums were plentiful on the trees growing amongst the rocks there. While he was hunting, he noticed an opening between two rocks. He put his foot in it and was drawn in. He found himself in one of the many caves in the vicinity. The cave was lit by a strange light, and was inhabited by many very small people who came to him showing signs of friendship. They called him Jambi, which is a general term for friendship, although it means brother-in-law. He tried to get back above the surface, but found that he had to wrestle with the little people. They were very strong, although small, and although he fought many of them, they all overcame him. Feeling exhausted he lay down to rest. The little people, the Nyols, gave him rugs to sleep on and grubs to eat. The latter were a great delicacy, and he enjoyed them very much. At last, many of the Nyols went away and he was left in the charge of one of them. Everything had been quiet, but now he heard a rustling sound. One of the Nyols came to him saying he would show him the way to the surface of the ground. Before very long he was amongst his own people, but for several days could not tell them what had happened to him. His mind had temporarily gone blank.

3.1.2 The Buchan District: a Brief History

In early 1838 Edward Bayliss travelled from Aston, Maneroo, in search of grazing country, through Suggan Buggan to Buchan where he camped on the river flat, marked out a temporary station, and determined he would return with cattle. Returning to Maneroo, his reports on the Buchan district were so positive, that John Wilkinson started immediately and took possession of Buchan. When Bayliss returned later in 1838, he found Wilkinson established in the Buchan district and he was forced to go upstream up the Buchan River valley where he established the Gelantipy run in 1839. Wilkinson and Bayliss only stayed two seasons in the Buchan and Gelantipy districts, moving to the south in search of new pastures. By 1842, Fraser Mowatt had acquired the Buchan run (BSC, 1989: 6). Archibald Macleod, and his sons Norman and John, took possession of the Buchan run in 1845, and remained in control until 1862. Leases for Buchan were transferred frequently (see Billis & Kenyon, 1974), until the run was forfeited in 1880.

The Buchan area homestead was surveyed in 1866 and amended the following year by contract surveyor Arthur F Walker. G.H. Wilmot, district surveyor, undertook a comprehensive survey of 8,000 acres in the parish of Buchan in 1868, however demand for land was limited, and the major part of the area was selected, and resurveyed at a later date. Buchan was proclaimed a town in May 1873, and the town boundaries surveyed in 1874.

The presence of large caverns at Buchan was first mentioned in Stuart Ryrie's 1840 report (Ryrie, 1840). *The Argus* (17/11/1854) published a report from a staff member of the office of the Surveyor-General employed on the Omeo goldfields, in which it is announced that gold had been found in the Buchan district. The report mentions that the caves in the Buchan district were not extensively known:

> We traversed a portion of the country on the Buchan River. ... Some of the caves, I am told, are very beautiful, but have not been as yet explored to any extent. In passing over parts covered with the most luxuriant grass, openings are to be observed, down which if a stone is rolled its re-echos can be heard for hundreds of feet. We visited some of the caves, but were unable to bring anything away, excepting a few small stalactites, &c., as we had no means of carrying them.

An early account of a visit to the Buchan caves was published in the *Gippsland Times* of 27 March 1873. The party stayed at the Buchan pastoral station where they received hospitality from the station manager as the nearest accommodation house was eight miles distant. The account provides details of the earliest known photography in the Buchan Caves.

> A start was then made for the Murrindal mine and Buchan caves. Mr Cornell accompanied us with his photographic apparatus, for the purpose of taking a view of the interior of the largest

> cave.[10] ... It is almost impossible to describe this wonderful place, it requires to be seen, and is well worthy of a trip from Melbourne. ... Mr Cornell succeeded in obtaining a photo of the pulpit rock, but not a good one, there is no doubt with proper appliances the whole interior could be taken (*Gippsland Times*, 27/3/1873).

In July 1873, Mr G.F. Ulrich, a geologist, in the company of a director of the Back Creek Silver and Lead Mine, visited Buchan, accompanied by a correspondent from Bairnsdale. This party also stayed at Buchan station (*Gippsland Times*, 5/7/1873). In April 1878 members of the government's Wattle Bark Commission, availed themselves of the opportunity of visiting the caves at Buchan (*Gippsland Times*, 15/4/1878).

On 28 August 1880, the *Australasian Sketcher with Pen and Pencil* published an article on the Buchan caves:

> DISTRICT SCENERY. THE BUCHAN CAVES. (BY E.M. Buchan.)
>
> The hills here are pierced with scores–nay probably hundreds, of caves. With the exception, however, of those contiguous to the township, but little is known of them. The cathedral cave is, not only one of the most interesting but the one best known, and I set forth to explore it. ... A lady friend, who visited this cave, informed me that one of her party was induced to sing here, and the effect of the harmony rolling along the vaulted roof and reverberating throughout the distant, and lofty intricacies of the cavern, was magnificent. A scientific friend also informs me, that entombed beneath the stalagmite, which forms the floor of the cave, the fossil bones of gigantic marsupials are sometimes to be found. ... I found the light from my lantern insufficient to enable me to see the roof of some of the chambers, and would suggest that visitors to these caves should provide themselves with magnesium wire. ... As it is, I think Buchan is destined to become a picturesque little hamlet well worth the tourist's visit, if he be anything of a mountaineer.

In 1885 a hotel named the Cricket Club Hotel was established in Buchan by Henry Schacht. In the *Gippsland Times* of 29 May 1885, Shacht informed its readers that 'tourists, travellers, and others will find his establishment replete with every convenience. The celebrated Buchan caves, which are well worthy of a visit, are within a short distance of the township, and visitors can always obtain a guide at the hotel. Saddle horses can be obtained on hire, and good accommodation paddocks are provided. Newport's coach from Bruthen also arrives and takes its departure from the hotel' (*Gippsland Times*, 29 May 1885). From that issue onwards, Schacht regularly promoted his services announcing in advertisements that 'Tourists and others desirous of visiting the Buchan Caves can be furnished with guides'.

Julian Thomas, aka 'The Vagabond', visited the Buchan district in May 1886 and reported on his experiences. At Bruthen he met Frank Welby, the owner of Buchan station, who served as his guide. Thomas is one of the first to mention the destruction of some of the stalagmites and stalactites by those who visited the caves.

---

10 F. Cornell, a photographer.

> The far-famed Buchan caves are largely composed of limestone, and consist of underground caverns, the extent and number of which is unknown. ... Provided with candles and kerosene torches, one follows the guide, who brings the visitor to "the pulpit, and chair, and canopy," formed of pure limestone, which sparkles beneath the light, casting weird shadows around. ... The Spring Creek caves and the Murindal caves are said to be equal to Wilson's caves. In some places these are highly dangerous, and one should not enter without a guide. Unfortunately, too, in all the caverns sacrilegious hands have broken off some of the finest stalactites and stalagmites (*The Argus*, 22/5/1886).

In 1889 the first geological survey of the Buchan Caves was commenced. James Stirling, Assistant Government Geologist with the Mines Department, published a description of Duke, O'Rourke, and Dickson (Dixon) caves, and the Spring Creek, Wilson Creek, and Murrindal caves. He recommended that the Buchan Caves be developed as a tourist attraction, along the lines of the Jenolan Caves in New South Wales. Stirling made ground plans of the Buchan and neighbouring caves and heliotype plates from the expedition photographs by J.H. Harvey, illustrating views in Wilson and Dickson caves. These photographs (and others by Harvey not published in the report) have long been seen as being the first photographs taken in caves in Victoria (Douglas, n.d); however it is clear that Douglas was unaware of the Cornell photographs taken in 1873.

In February 1889, Stirling delivered a lecture entitled 'The Caves of Australia' to the Sale Branch of the Australian Natives Association. *The Gippsland Times* (25/2/1889) published an account of his lecture:

> The Buchan caves were of great interest to Victoria at the present time. Although known for many years, no systematic attempt was made to explore them until recently when his assistants, Miners Ralston and Tetu, accompanied by Mr Rellie, of Buchan, began to make a survey of the chambers, under his instructions. Having visited the caves, and examined the newly discovered chamber, he could with safety say that they were certainly the best caves yet discovered in Victoria. Although, so far as is known at present, they were not as large as the Jenolan caves, yet in many respects they were quite equal to these now celebrated caverns in point of beauty and interest. ... After tracing some of the steps in the course of time revealed by the caves, the lecturer concluded by expressing a hope that some of his fellow Australian Natives would be roused to the importance of investigating the underground wonderland of Australia, and, by searching, discover many relies of bygones ages in this great country which supplied so many natural history curiosities to the world's museums. ... At short intervals Mr Harvey, secretary to the Photographic Association, Melbourne, who is accompanying Mr Stirling, exhibited views by the aid of a magic lantern, illustrative of the lecture.

On 1 May 1889, the *Bendigo Advertiser* reported that acting on Stirling's recommendation, 'the Minister of Mines has recommended the Lands department to permanently reserve the celebrated Buchan caves which Mr. Sterling [sic] so graphically describes'. The *Bairnsdale Advertiser and Tambo and Omeo Chronicle* (2/5/1889) explained that Stirling had recommended their permanent reservation 'otherwise many of the striking physical and natural characteristics of the colony may otherwise lose their freshness and scientific value'. *The Gippsland Times* (12/3/1890) quoted from Stirling's report:

> The widespread interest which the Buchan caves undoubtedly obtain, ... should, I think, suggest the advisability of having a proper supervision and further scientific examination made. I would recommend that, in addition to the reservations which have now been made at Spring Creek, Wilson's Cave, and Dixon's cave, some competent person should be appointed as care-taker, whose duties it would be to complete the exploration for new chambers, construct ladders and hand rails in different parts of the caves, and systematically carry on the work of removing portions of the stalagmital floors in search of fossils. The specimens already obtained in the breccia and cave earth being deemed of sufficient importance to warrant further scientific explorations, should it be desired to light up the caves by electricity, especially the Royal chamber in Wilson's Cave, there should be no difficulty, as water power to drive a dynamo is readily obtainable from the Buchan River close by.

In 1897, the *Bairnsdale Advertiser and Tambo and Omeo Chronicle* reported on the state of research at the Buchan Caves:

> The great caves at Buchan, and the scenery in their neighborhood–the Devil's Glen, the Snowy River's junction with the Buchan, and the Pyramids, for example are well worthy alike the attention of the geologist and the tourist, the scientist and the mere lover of the beautiful in nature. ... Considered geologically the formation of rocks in the Buchan valley belong to the middle Devonian period. The caves are full of stalactites and stalagmites, the largest chamber being about 100 feet long by 35 feet wide, and 35 feet high. As yet no successful search has been made for fossil remains of the gigantic kangaroo, or mammoth wombat, so there is a great field yet untouched for subterranean exploration. The three principal caves at Buchan are Wilson's cave, distant about 4 miles from the Buchan post office; the Spring Creek cave, about one mile; and the Basin Creek cave, about 14 miles to the north east. There are number of other caves in the district, but not containing many stalactites. It is generally believed by residents of the district that there are other caves yet to be discovered, but it is very difficult to find the mouth or entrance to a new cave, as it is usually on the side or at the foot of a hill, and covered with grass and ferns. ... None of the principal caves have been thoroughly explored, and it is reasonable to suppose that there are other chambers yet undiscovered, awaiting in their awful darkness for someone to enter them and behold their beauty (*Bairnsdale Advertiser and Tambo and Omeo Chronicle*, 10/8/1897).

A.E. Kitson, a geologist with the Mines Department, in 1900 reported on the caves along Spring Creek. He recommended that cave reservations be set apart along Spring and Cave (now Fairy) creeks, at Dickson, Slocombe, and Wilson caves, in the vicinity of The Pyramids and at the Camping Reserve south of the Dickson Cave area. By 1900 many of the more accessible portions of the caves had been damaged by vandalism, but Kitson recommended that outstanding features should be preserved and suggested new passages and chambers would likely be found along unexplored portions of the cave complex. As a result of Kitson's report 65 ha, being the unsold portion of the Buchan township, were set aside as a caves reserve by the Department of Crown Lands and Survey (Government Gazette, 19/7/1901; Swift, 1951: 3), and 48 ha adjoining this in the vicinity of the Spring Creek caves were also reserved (Government Gazette, 29/1/1902).

In January 1902, the Bairnsdale District Commercial and Progress Association discussed the Buchan Caves during one of their monthly meetings. Members expressed concern that the Government was not taking any care of the caves, and there were calls for their protection. The installation of electricity was recommended as well as the development of the caves for tourism (*Bairnsdale Advertiser and Tambo and Omeo Chronicle*, 18/1/1902).

In April 1902, 'Vigilance', a concerned citizen from Bairnsdale, wrote to *The Argus* (10/4/1902) about the protection and preservation of the Buchan Caves:

> The Spring Creek caves, being most accessible, have already suffered severely from the wanton acts of vandalism perpetrated from time to time by various visitors some of what were doubtless the most beautiful of the chambers therein having been ruthlessly destroyed and hundreds of stalactites broken off and removed. In consequence of the want of supervision these wonderful phenomena which were the result of Nature's operations through countless ages, and which should have been the delight of generations to come, have been shorn of much of their attraction, and no punishment could be too severe for those who have so shamefully demolished them. In like manner the other caves in the district of Buchan are being ruined, and unless measures are at once taken to protect them their beauty will be gone, and what should be the means of attracting increasing numbers of tourists from this and other states to what is one of the most healthy localities in Victoria will entirely lose their charms. With the example of New South Wales before us it is surely not too much to ask that caretakers be appointed for the various caves and better facilities for reaching and exploring them be given, such as tracks to the month of the caves and ladders to descend to the various chambers. Were such methods of popularising them adopted there would, I am sure, set in to these beautiful spots a regular stream of visitors equal in magnitude to that which now wends its way to the Jenolan and other caves in the neighbouring state. They only require to be known more widely to be more fully appreciated.

In April 1902, the *Bairnsdale Advertiser and Tambo and Omeo Chronicle* published a series of articles on Buchan and its Caves by 'Mascotte'.

> The caves have been almost completely stripped–denuded of almost the last stalactite. And you are told, and can easily believe–that the others will soon be in a like deplorable condition unless a somnolent, listless Government wakes up once in a while and hires a man to watch these caves with a good deal of carefulness and a gun, but that is quite unlikely. ... But all the caves are not destroyed, by any means (*Bairnsdale Advertiser and Tambo and Omeo Chronicle*, 15/4/1902).

Mascotte's second article concerned the Spring Creek caves (*Bairnsdale Advertiser and Tambo and Omeo Chronicle*, 19/4/1902). The third article was about the Basin caves. The third article is of interest as it shows the actions of a local resident who took it upon himself to provide a level of protection to the caves near his residence.

> The Basin caves are some distance from the township–about 12 miles, in fact–and the road thither is mostly uphill, and tedious. Up till quite recently these caves were much the best preserved of all, principally due, no doubt to their distance from the town and the difficulty of

> access. Lately, however, wholesale depredations have been made upon their contents by systematic robbers who, despising such commonplace means of despoilation as are practised by everyday vandals actually went to the caves equipped with a horse and cart, and wrecked a great proportion of them to fill the vehicle. A neighbouring resident Mr Slocombe, has wasted a considerable amount of his own time in writing to the Government with respect to the ruthless destruction that has been going on and asking that some steps ought be taken to put a stop to it. But this being a purely national matter and not one of merely political expediency or personal profit, no notice has been taken of either the representations or the request, except that a few notices, warning visitors not to interfere with the caves, have been posted up. Ultimately Mr Slocombe took matters into his own hands and fastened up the entrances to several of the best caves, and now entrance to these cannot be obtained without his consent–which is a very good thing. But the preservation of these caves is not a matter for unremunerated private action. It is the unmistakable duty of the Government to preserve such remarkable curiosities as they represent, at whatever cost (*Bairnsdale Advertiser and Tambo and Omeo Chronicle,* 29/4/1902).

The *Bairnsdale Advertiser and Tambo and Omeo Chronicle* (26/8/1905) continued to maintain the pressure for government action in an editorial published in August 1905:

> Successive Government geologists have urged that some proper steps should be taken to "develop" by exploration the beautiful limestone caves at Buchan. The present occupant of the position Mr Dunn, has furnished the Government with a lengthy report on the caves, and a recommendation that cash rewards should be offered for the discovery of new caves. He also recommends that all possible precautions should be taken to protect the caves from vandalism. ... There are very few people living in the vicinity of the caves, and visitors are exceedingly infrequent. ... Mr Dunn considers it as exceedingly probable that a big national asset is going to waste at Buchan, and he fully believe that proper exploration work will lead to very important discoveries. ... If it was the particular business of any individuals to attract visitors to Buchan the trip there would become one of the most popular on the continent, for it would possess attractions that are not to be found anywhere else. But at present it is no-one's business. The caves are to a great extent neglected because almost wholly unknown outside the district. But for the zealous and unremunerated care of the cave man who takes any real interest in them and who for years officiated as their caretaker, the caves would long ago have ceased to be in any way attractive. Mr Dunn's suggestion for the offering of cash rewards for the discovery of new caves is a good one, no doubt, but its ultimate value must depend upon whether or not the Government take steps to have all the caves that have been or may yet be discovered "reserved" and properly cared for. Someone must be appointed to look after them and must be paid for doing it, of course. Everything in connection with the development and protection of the caves must be done by the Government, because there is no public body or community of people to whom the responsibility could be delegated. The municipality of Tambo, in the territory of which the limestone deposits are situated, is not interested in caves, and any attempt to get it interested to a sufficient extent to undertake the work of protecting the known caverns and prospecting for more would he hopeless at the outset. The question is one wholly for the Government and properly so, seeing that the whole matter is a national one. It is very much to be hoped, from every point of view, that the recommendations that the geological director has made to the Government in connection with the caves at Buchan will be acted upon and amplified (*Bairnsdale Advertiser and Tambo and Omeo Chronicle*, 26/8/1905).

3.2 Second Phase: Framing and Elevation 1907–Present

The second phase identified by MacCannell (1976) in the evolution of attractions is 'framing and elevation' which he argued results from an increase in visitation, when demand requires some form of management intervention, whereby the sight is displayed more prominently and framed off.

One local resident who responded to the challenge to explore the Buchan cave complex was a young athlete and local resident named Francis (Frank) Herbert Arthur Moon. *The Argus* (3/10/1906) reported on one of his discoveries:

> BUCHAN, Tuesday–The new cave discovered by Frank Moon, a young athlete, is situated on the south side of Spring Creek, half a mile from the Buchan Post-office. Moon noticed a fissure in the rock, and determined to investigate. With three comrades, he entered a narrow fissure, measuring 6ft. by 2 ft. with acetlyne [sic] lamps. ... The cave is absolutely perfect, and out-rivals all others in the district in extent and rugged beauty. It can be easily drained (*The Argus*, 3/10/1906).

As a consequence of Moon's discovery, the Victorian Premier, Thomas Bent, despatched A.E. Kitson, an officer of the geological branch of the Mines department to investigate and report. According to *The Argus* (11/10/1906), 'Certainly the earlier discovered caves at Buchan have suffered severely during the past 10 years from wonton acts of vandalism, but there are still left immense subterranean areas, which should prove invaluable as attractions for tourists. The unaccessability [sic] of Buchan has been the only factor keeping these caves back for so many years. When once the township is reached the caves are close at hand. The most recently discovered is entered on a chamber only half a mile from Buchan Post-office. But Buchan itself is remote from any railway line, and at present is difficult of access. The work of providing means of reaching these caves will be given to the Tourist Bureau which it is proposed to establish' (*The Argus*, 11/10/1906).

The Gippsland Times reported on the findings from Kitson's investigation:

> Mr. Kitson has paid several visits to these caves and he reports that there are several known groups, in addition to the one discovered a few days ago. They are as follows: Spring Creek Caves, Buchan; O'Rourke's caves, Buchan; Green caves, Buchan; Slocombe's caves, on Basin Creek, 12 miles north-east of Buchan; Murrindal caves, six miles north-east of Buchan. ... Mr Kitson reports that the caves have never been systematically explored, and he has no doubt whatever that careful examination and labor will reveal numerous additional chambers, adorned with stalactites, stalagmites and stalactical drapery. In several portions of the explored passages there are small openings which probably lead to larger passages and chambers. As a result of representations formerly made by Mr Kitson, 120 acres of the mining reserve and 160 acres of the township reserve at Bruthen have been reserved for the preservation of caves. Spring Creek runs into the "Bruthen River just outside the town, and a small tributary called Cave Creek runs into Spring Creek, the caves on these waterways being close to the town. The Spring Creek, Wilson's, Dickson's, Green and Basin Creek caves, says Mr Kitson, are particularly worthy of preservation, for though all but the last have suffered from the vandalism of sight-seers, who have destroyed most of the smaller stalagmites and stalactical drapery within reach, there still remains a wealth

> of natural beauty, besides which there is a great probability of the discovery of new caves and chambers. In advocating the preservation of these limestone caves, Mr. Kitson says that the limestone country generally in the Buchan district has a very pleasing aspect. ... With improved roads to Buchan, and the opening up of the caves, Mr Kitson is confident that Buchan will become a celebrated tourist resort (*Gippsland Times*, 15/10/1906).

As a consequence of this new discovery, the Minister of Lands instructed an officer of his department to ascertain how far the caves underlie Crown land. Once this information was obtained, it was the minister's intention to authorise a local resident to act as caretaker for this portion of the caves (*Portland Guardian*, 22/10/1906).

The Argus continued to agitate for the immediate protection of the Buchan Caves. 'Local residents are already exhibiting stalactites and stalagmites taken from the new cave, and, owing to the use of the "emergency torch," a piece of bark saturated in kerosene, some beautiful formations have been destroyed' (*The Argus*, 25/10/1906). Frank Moon was appointed caretaker and he threw himself into the task of exploring the Buchan Caves complex.

The 'Fairy Cave' was 'discovered' by Frank Moon, on 16 March 1907 (*Bairnsdale Advertiser and Tambo and Omeo Chronicle*, 19/3/1907). He saw a small hole or crevice in the side of a hill, and enlarged it with some gelignite, and descended fifty feet to what is now known as 'Fairy Cave' (Salierno, 1987: 55). Paths were constructed through the cave and wire netting installed to protect the decorations. Fairy Cave was opened to the public in December 1907. Initially, lighting was provided by candles given to the visitors and magnesium lamps used by the guides. Electricity was connected in 1920 when a generating plant was installed, and this was used until 1969 when State Electricity Commission power was connected. Subsequently there has been much upgrading of the Caves and associated infrastructure.

In April 1907, W. Thorn, of the Lands department, visited the Buchan Caves to report to the Minister on the recent discoveries and their comparison with the Jenolan Caves, in New South Wales, of which he recently made an inspection. He also reported on any works necessary for opening up the district to tourists and for the protection of the caves. It was hoped that the new caves will be opened up sufficiently to attract a considerable number of visitors next season (*The Argus*, 5/4/1907). The *Bairnsdale Advertiser and Tambo and Omeo Chronicle* reported that the Buchan caves were becoming better known. 'Recent visitors who have returned to the metropolis speak in glowing terms of the wonderful sights they saw in their inspection of these new caves. This should have a good effect in hurrying on the department to open up the roads to Buchan and making the caves accessible to the public. It is unlikely, though, according to a statement made by the Minister of Lands to day, that much can be done in this direction at present. Mr Mackey has promised, however, to have the arrangements for rewiring the caves completed by next summer. The roadwork will probably be finished a little sooner' (*Bairnsdale Advertiser and Tambo and Omeo Chronicle*, 4/4/1907).

The 'Royal Cave' was located in November 1910, by a party led by Frederick J Wilson, Caves Supervisor of the Buchan Reserve since 1907, and the discoverer of the Jenolan Caves in New South Wales (Swift, 1951: 6). The others were Frank Moon and Constable Brown, a local policeman (Allen correspondence 28/5/1968 in File 'Caves and Tourism'). *The Argus* (9/11/1910), in reporting its discovery, noted that it 'is said by experts to equal if not excel, anything at Jenolan'. At Royal Cave, in 1913, William Bonwick and William Foster, reserve employees, cut through a solid block of black marble, and used a large quantity of explosives to get through 150 feet. The cave opened to the public in November 1913.

The Argus in 1951 summarized some of those responsible for the framing and elevation of the Buchan Caves:

> Associated with the opening and exploration of the caves was the late Dr. John Flynn, later famous as Flynn of the Inland. The Doctor was one of the first to enter the new chambers, and his photographs of them, made into lantern slides, were shown to big audiences. He convinced the Government of the day that the Buchan caves were a wonderful treasure. Parliament acted, and Frank Moon was asked to take charge of the caves and explore the unknown portions. His first important discovery came on March 18, 1907, when he entered the Fairy cave from a fissure on the hillside. From a side passage in its depths he discovered the Royal cave a year later, but falls of rock made the way unsafe, and a new entrance tunnel was driven through the hill. Before the visitors could come to marvel at this fascinating underworld there was hard work to be done. WITH the guidance of Mr. F. Wilson, the discoverer of Jenolan, the tunnels, passageways, and illuminations were planned and commenced. Others played their parts, too. The late William Bonwick and the late William Foster, before the days of the pneumatic drill, wielded hammer and chisel day after day, clearing passages through the solid rock. Cement for the later tunnels and stairways was carried in bucketful by bucketful by Frank Moon's son-in-law, Frank Hansford. DAVID SWIFT (*The Argus* 16/11/1951).

In July 1918, a committee of Management was constituted which continued until its reconstitution as an advisory committee in 1946. In October 1918, *The Argus* reported that the Minister of Lands was concerned that the caves were now illuminated with magnesium ribbon which was rapidly destroying the finer features of the caves, and proposed to install electric lighting. 'The cost of the magnesium ribbon light was £50 a year. Last year 3,465 tourists visited the caves and their admission money represented a revenue of £350. In addition revenue had been derived by the railways. For £670 the caves would be electrically lit and there would be no recurring expense as was the case now' (*The Argus*, 11/10/1918).

In April 1925, *The Northern Miner* noted that 'The thousands, of people who have already seen the Buchan Caves, and the tens of thousands who will see them in the future, will be thankful that this invaluable sight-seeing resort was saved to the State in the nick of time ... But for the strong and persistent recommendations of geologists, who were able to realise their worth, most of the area embracing the caverns would have passed from the Crown into private hands 25 years ago' (*The Northern Miner*, 21/4/1925). Yet *The Argus* was critical that the Victorian Government was not doing enough to promote the Buchan Caves. It considered that 'If they had been the property of private individuals instead of the State they would have been earning a fortune long ago' (*The Argus*, 5/2/1926).

3.3 Third Phase: Enshrinement 1937–Present

MacCannell (1976) has identified 'enshrinement' as the third phase in the evolution of attractions when framing material that is used has itself entered the first stage of sacralization (marking). In the case of the Buchan Caves this would most likely refer to the facilities and other activities that are available to the visitor at the Caves, such as picnicking and camping, activities that don't necessarily involve any interaction with the nucleus that is the caves themselves, other than having them provide the location or setting for the activity.

In 1937 *The Argus* reported that a development plan for the Buchan Caves was being prepared by the committee of management of the resort. 'The proposals to be developed include the establishment of native game sanctuaries and a national park, the provision of camping facilities for at least 300 pic-nic parties, a swimming pool and the cutting of tracks through about 700 acres of bush. Mr Lind said last night that he hoped that the resort would become one of the most attractive in Vicoria. If the estimates of cost are approved the work will be undertaken soon by the Public Works Department' (*The Argus*, 24/11/1937). The following year the *Gippsland Times* (19/5/1938) reported that the improvements being made at Buchan Caves National Park were coming from unemployment relief funds. The works included 'a swimming pool fed by the underground river, camping and parking areas, children's playground, and tennis courts. The reserve is to be made more attractive by extension of the tree-planting scheme, including an inner reserve in the lower part of the valley for native fauna, Proceeds from admission charges to the caves are paid into consolidated revenue.The chairman of the caves committee is the chief clerk of the Lands department' (*Gippsland Times*, 19/5/1938).

The Buchan Caves National Park was formally dedicated on 3 December 1938 by the Minister for Forests (Mr. Lind). 'Mr. Lind will also open officially the new automobile camping park in the Buchan Caves National Park at 3 p.m. The camping park is claimed to be the most modern in Australia. It has facilities for dances and concerts, a kitchen, and dining hall, showers, bathrooms, a swimming-pool sunk in marble rock, tennis-courts, children's playground, fire-places, car-washes, and it is sewered. An all-weather road leads to it from the Prince's Highway at Nowa Nowa' (*The Argus*, 2/12/1938).

The Caves were closed in February 1942 because all the staff had enlisted for military service in WW2, and owing to the difficulty of maintaining transport facilities between Lakes Entrance. In April 1946, the Caves were re-opened when two members of the former staff were demobilised from the army, and after some reconditioning and overhauling of the electric light and pumping plant at the cave complex had been undertaken (*Gippsland Times*, 10/1/1946).

3.3.1 Management History of the Buchan Caves

AE Kitson, a geologist with the Mines Department, in 1900 reported on the caves along Spring Creek. He recommended that cave reservations be set apart along Spring and Cave (now Fairy) creeks, at Dickson, Slocombe, and Wilson caves, in the vicinity of The Pyramids and at the Camping Reserve south of the Dickson Cave area.

By 1900 many of the more accessible portions of the caves had been damaged by vandalism, but Kitson recommended that outstanding features should be preserved and suggested new passages and chambers would likely be found along unexplored portions of the cave complex. As a result of Kitson's report 65 ha, being the unsold portion of Buchan township, were set aside as a caves reserve by the Department of Crown Lands and Survey (Swift, 1951: 3), and 48 ha adjoining this in the vicinity of the Spring Creek caves were also reserved. Yet, the Caves were not formally opened to the public until 2 December 1907. Initially, lighting was provided by the guides who carried lanterns and magnesium torches or flares. Electricity was connected in 1920 when a generating plant was installed, and this was used until 1969 when State Electricity Commission power was connected. Subsequently there has been much upgrading of the Caves and ancillary facilities. Wire netting was installed to protect the formations, and over the years obstructions have been removed, and board walks and concrete paths constructed to facilitate public inspection. In June 1907, Frederick Wilson was appointed Caves Supervisor. In July 1918, a committee of Management was constituted which continued until its reconstitution in 1946.

The camping reserve at Buchan Caves was proclaimed in 1930, and in August, Hugh Linaker was commissioned to landscape the reserve. In December 1938, the Buchan Caves National Park was officially opened. According to the article 'Buchan Caves Attract the Tourist', in the 1975 edition of *Landmark*, the journal of the Department of Crown Lands and Survey, the Buchan Reserve comprises some 280 hectares. Aitken (1994) claims 285 hectares. In May 1939, a new electricity generating plant was installed at the the Buchan Caves reserve at a cost of some £1,473/7/7 (*The Argus*, 13/5/1939).

In 1946, an Advisory Committee, under the chairmanship of EJ Pemberton, was formed to advise the Department of Crown Lands and Survey, on the development, maintenance and supervision of the caves. In 1951, the advisory committee comprised representatives from the Departments of Lands and Survey; Public Works; Victorian Railways, and a Lands Officer from Bairnsdale (Swift, 1951: 6).

Boadle (1991) produced a draft management plan for the karst and cave resources in the Buchan and Murrindal area. In that plan, it was stated that the Buchan Caves Reserves consist of nine separate blocks, the largest being immediately west of the

Township of Buchan, generally known as the 'Buchan Caves Reserve'. Boadle (1991: 44) recognized the important role played by visitor facilities and services at the Main Reserve, and a draft management plan highlighted the need for information provided to be of a high standard, and recommended the production of a high quality brochure for tourists. The plan acknowledged that the Main Reserve was a focus of tourist activity, and an educational centre visited by many secondary and tertiary groups.

In 1994, Richard Aitken prepared a Classification Report for the National Trust of Australia (Victoria) of the Buchan Caves Reserve. Aitken considered the Buchan Reserve ranked second to the Jenolan Caves Reserve in New South Wales, in terms of its contemporary popularity, impact on local and regional development, surviving attributes and historical importance. He noted that the Buchan Reserve contains many fine examples of protective features, and the period and nature of the tourist development is comparable with places such as Mt Buffalo Chalet. A notable feature of the Buchan Caves Reserve is the degree to which it has been modelled on the United States of America National Parks Service, particularly the adoption of 'parkitecture' styles for some of its buildings.

In 1995, the Buchan Caves Committee of Management received a grant of $6,000 from the National Estates Grant Program. The project involved the production of an interpretative display located at the Buchan Caves Reserve, focussing on the significance of Aboriginal culture in the Buchan area, particularly Cloggs Cave (Calnin, 1997). It was emphasized in the application that the Aboriginal heritage of the Buchan district was something the visitors to the Buchan Caves complex were 'not normally exposed to'. The 'Cloggs Cave interpretation sign' was installed at the Buchan Caves Reserve in August 1996. The text of the sign is presented in Appendix 3.1.

3.4 Fourth Phase: Duplication

MacCannell's (1976) fourth phase in the development of attractions is that of 'duplication', when copies of the nucleus of the attraction, in this case the caves, are made available through media such as paintings, photographs, and postcards.

The first photographic reproductions of scenery in the Buchan Caves district were taken in early 1873 by a Mr. F. Cornell. 'The series includes views of the Buchan Station, the workings of the Buchan Lead and Silver Mine, Back Creek, and of an exquisite little waterfall immediately adjacent thereto; also a view of the entrance to the celebrated Murrindal Cave, and one actually taken within the cave, which represents the pulpit rock' (*Gippsland Times*, 25/2/1873).

At the sixth annual exhibition of the Amateur Photographic Association, held at the Athenaeum in Melbourne, Mr. J.H Harvey, the Secretary of the Society, had a half dozen views of the Buchan Caves taken by the magnesium riband (see Fig.3.3). 'The illustrations were extremely well executed, the various parts of interest, the stalactites, the stalagmites, the side chambers of the caverns, the limestone formations being delineated with exactness' (*North Melbourne Advertiser*, 19/9/1890).

The *Bairnsdale Advertiser and Tambo and Omeo Chronicle* (10/8/1897) reported that some fine views of the interior of the Buchan Caves and of the picturesque spots in their vicinity had been taken by Messrs Ward Bros., of Bairnsdale.

In 1906, the *Bairnsdale Advertiser and Tambo and Omeo Chronicle* included an interior scene of one of the Buchan Caves in a series of 35 "postcards" of views now on sale at *The Advertiser* office for 9d a dozen (*Bairnsdale Advertiser and Tambo and Omeo Chronicle*, 26/4/1906). The following year, "Views of the Buchan Caves and Pyramids," was compiled by F. Verrell Heath, and published by T.C. Lothian, Melbourne (*Gippsland Times*, 11/4/1907). The same year (1907), F. Whitcomb released *Guide to Buchan Caves and Gippsland Lakes*, and included photographs by H.D. Bulmer and N.J. Caire and was published by the Cunninghame Progressive Association.

In 1926, *The Argus* reported that C.H. Herschell had taken films of the interior of the Buchan Caves. 'The caves have never been filmed before, and magnesium flares of the kind used to make motion pictures of forests by night in America were fired' (*The Argus*, 5/2/1926).

3.5 Fifth Phase: Social Reproduction

For MacCannell (1976), the final stage of sight sacralization is social reproduction which occurs when groups, cities, and regions begin to name themselves after famous attractions. In the case of Buchan, the name is found in the Buchan pastoral run, first taken up in early 1839, Buchan River, and Buchan township. In 1935 the Buchan Hotel was changed to the Caves Hotel (*The Argus*, 18/6/1935). Since 1935 the hotel's name was changed to Buchan Caves Hotel, its current name. It is also possible to identify numerous business that include Buchan in their business names, such as Buchan Valley Log Cabins; Buchan Lodge; Buchan Motel; Buchan Cottage,

3.6 Tourism at Buchan Caves

The first mention in tourism literature of tourism at the caves dates from *Bailliere's Victorian Gazetteer and Road Guide* of 1879 where mention is made of caves near Buchan (Aitken, 1994). Similar brief references to the 'famous Buchan caves' appeared in Pickersgill (1885). However, the earliest known reference to tourist use of the caves is believed to be R.S. Browne's (1886) *Our Guide to the Gippsland Lakes*. In July 1887, the *Bairnsdale Advertiser and Tambo and Omeo Chronicle* notified its readers that Edward Foley, the local mail contractor, had placed a coach on the line between Buchan and Bruthen and those desirous of visiting the far-famed Buchan caves were able to make reservations with the contractor.

The *Bairnsdale Advertiser and Tambo and Omeo Chronicle* (4/3/1902) reported on a plan by the proprietor of Lake Tyers House to link tourism at Lake Tyers with that at Buchan.

The Argus summarized the state of tourism to the caves in 1914:

> As evidencing their growing popularity of the Buchan Caves now recognised as one of the 'show places' of Victoria, the increase in the number of visitors this year is very satisfactory. During January 370 admission fees were paid and during February 251. Since the caves were opened six years ago the total admissions have numbered about 8,000. In the immediate vicinity of Buchan there are now four caves open-Fairy, Blackwood, Royal and Moon Caves. By the opening of the Royal Cave during this tourist season some of the most beautiful chambers containing exquisite limestone formations have been made available to visitors. ... Many of the visitors consider these formations rival those of the Jenolan Caves. By the recent developments the value of Buchan as a tourist resort has been very much enhanced and the extent of cave passages now fully opened up and easily accessible will provide ample enjoyment for visitors for many hours. It is no longer possible for visitors to view the caves as they should be seen by merely staying in the township one night. Tourists should arrange to devote several days to their inspection (*The Argus*, 17/3/1914).

Following from this report, *The Argus* published letters from two correspondents who bemoaned the fact that although they had purchased tickets they were constrained by the limited number of conveyances to take them to the Buchan Caves from Cunninghame. They were also critical of the state of the road from Nowa Nowa to Buchan. The first correspondent, from Kyneton, closed his letter stating 'Present conditions will not conduce to the desire for a second visit' (*The Argus*, 21/3/1914). The second correspondent, from Geelong, agreed with the Kyneton writer (*The Argus*, 24/3/1914).

The Malvern Standard (26/4/1919) reported that year by year 'the number of tourists who visit the Buchan Caves is increasing. The summer months are the favourite times for making the visit, which is well repaid by the splendid scenery that abounds en route, not to speak of the pleasure derived from an inspection of the caves. Buchan is reached in one day from Melbourne by taking train from the central station to Nowa Nowa, at the head of Lake Tyers, where motorcars are in readiness to take tourists to their destination'.

In February 1922, *The Daily News* reported on the state of tourism at the Buchan Caves:

> These caves, taken under control only a few years ago, have been so highly developed, and opened up by safe, clean, easily traversed and airy passages, that, without effort and under conditions of comfort, young and old may now explore these wonders of nature which, seen under the bright electric light, or the still brighter blaze of the magnesium light, have delighted tens of thousands of visitors, and annually attract thousands to this fascinating resort. ... A great error is made by hundreds of tourists, who are led to believe that the caves can be inspected in an hour or two, and plan their tour accordingly for just a flying visit. Considering the distance to be travelled, and the expense involved, this should be avoided. As the length of the passages opened up for the convenience of visitors is over one and a half mile, it is very evident that considerable

time should be given for a reasonable and satisfactory examination. Again, a popular idea is that to see one cave is to see all. This is another decided mistake for, strange though it may appear, and incapable of explanation, all the caves differ very much in their formations. True, stalactites and stalagmites are in all, but in shape, color, and variety of form, they are distinctly dissimilar. A proper inspection of any of the principal caves will occupy from 1½ to 2 hours, and then it is quite time for the visitor to go out into the open air, when he may enjoy the freshness and charm of the park-like reserve, which now has a very fine avenue of eucalypts, acacias and other varieties of trees as a means of approach to the different cave entrances. Kangaroo and emu have been introduced into the reserve, and at times these are to be seen along the avenue. The principal caves in the Buchan Park are Fairy, Blackwood, Royal, Federal and Moon, the first discovery being the Fairy. ... In the Buchan Caves the State of Victoria has a natural asset of incalculable value, one that is worthy of the utmost care and the highest standard of development (*The Daily News*, 21/2/1922)

In January 1924, Frank Moon, the caretaker, discussed entrance fees and other grievances from tourists and gave some inkling of visitor satisfaction from visiting the Buchan Caves:

Sir,–As caretaker and guide to the Buchan Caves, I desire to reply to recent letters on charges for admission. The caves are open for inspection on all week days at 10.30 a.m., 2.30 p.m., and 7.30 p.m. The admission to the two principal caves–Fairy and Royal–is 2/6 for each person. Mr Shaw is correct in saying that people are charged 3/- for the evening visit. Half the above fees are charged for persons under 14 years of age. For other caves of less importance the charges are 1/6 and 1/-. Visitors frequently attack me about the bad state of the Ironstone crossing and Flourbag Hill between Nowa Nowa to Buchan. The continual "nagging" accounts for the guide 'holding forth' to that memorable assemblage referred to by Mr Shaw, who says that I stood on an eminence. As a matter of fact, I was standing alongside "Lot's Wife," and she did not even blush. I told those present to air their grievances through the press. Speaking with an unbiased mind, I find it impossible to make any comment or comparison between Buchan and Jenolan cave systems. Both are extremely beautiful and instructive. I would strongly advise any person wishing to have "a cheap 2/6 worth" to visit these beautiful Buchan Caves, and see for themselves. The usual comment from thousands of visitors is thus–Worth a 'quid' of any man's money. Good-bye old chap. I will see you again at Easter. I am going to bring my friends "-Yours &c. Buchan; Jan. 15. FRANK H.A. MOON (*The Argus*, 18/1/1924).

In 1926, the Swanston Motor Tourist Bureau was offering tours, inclusive of accommodation, to the Australian Alps, Mount Buffalo, Omeo, Gippsland Lakes, and Buchan (8 days, cost £12.10s), and to the Gippsland Lakes and the Buchan Caves (8 days, £9.10s) (Wells, 1986: 258).

In 1936, Winifred Moore published the following account of a guided tour of the Buchan Caves: she waxed lyrically of the role of the guide, Frank Moon:

As one who was instrumental in discovering and helping to develop the caves, our guide, a man named Moon, had a personal interest in his job,and having only our small party to conduct he was in fine form. Though he has been showing people through these caves for the greater part of his life — interrupted only by a break while on active service in France— familiarity has not dulled his enthusiasm. His wife and three daughters – apparently share this also, for with great pride he pointed out the very place in the cave in which one of his daughters—'Fairy Moon'–

> had actually been married to one of the younger guides. He himself, he said, had wished to be married in Fairy Cave, but the Government of the time, less broad-minded than that of to-day, had refused permission for the ceremony in such a place. ... So enthusiastic was our guide that as we splashed through the stream of seepage water he would call a halt now and then, and manipulating the red and white lights at a nearby switchboard, would direct our attention to the beautiful reflections made by the grottoes in the muddy stream at our feet. ... With his colleagues he had shown more than 600 visitors through the various caves on one day early in January, and even at the time of our visit, in the off season, he was busy explaining all their marvels to an average of between 60 and 70 tourists daily (*The Courier-Mail*, 20/2/1936).

In 1988, a Far East Gippsland Tourism Strategy was produced and Buchan was identified as a 'tourism hub'.

Visitation through the show caves during the decade 1981-91 averaged 85,082 people per year, with approximately 160,000 visiting the Main Reserve each year (Boadle, 1991: 41).

The Bataluk Cultural Trail is a collaborative project between five Aboriginal community organizations in East Gippsland and local government, concerned with cultural and eco-tourism. The trail integrated 12 separate cultural sites, stretching from Sale to Cann River, and included the Buchan Caves. Brochures were published and distributed from May 1995 (East Gippsland Institute of TAFE, 1996: 13). The trail was officially launched on 27 October 1995.

In October 1996, Urban Spatial and Economic Consultants Pty Ltd (USEC) completed an audit of the tourism infrastructure of Gippsland. They identified the Bataluk Cultural Trail had state-wide significance (USEC, 1996: iii). The Bataluk project was prioritized as it offered the visitor to Gippsland a unique Aboriginal perspective. They considered the trail required minimal capital input and provided opportunity and flow on effects for the Aboriginal population of the region (USEC, 1996: 22). In an analysis of the trail, USEC (1996: 85) claimed that Buchan was 'where aborigines would retreat for the winter months and live in caves'. In 1997, Urban Spatial and Economic Consultants Pty Ltd (USEC) produced a regional tourism development plan for the Lakes and Wilderness tourism region. In terms of Indigenous issues, the plan (USEC, 1997: 10) acknowledged that 'Lake Tyers and Buchan were significant current Aboriginal tourism areas'. Aboriginal tourism, via the Bataluk Cultural Trail, was identified as a product strength of the region. The development plan recommended the development of the Snowy River as a 'heritage icon'; development of the Buchan township and Buchan Caves as a heritage precinct for the region; and further expansion of the Bataluk Cultural Trail (USEC, 1997: 40).

In December 1997, the Department of Natural Resources and Environment and Parks Victoria (NRE & PV) produced an ecotourism strategy for Far East Gippsland, for Parks Victoria and the Department of Natural Resources and Environment. It was found that approximately 472,000 people annually visited public land in East Gippsland, 332,000 to parks and reserves, and 140,000 to state forests (NRE & PV, 1997: 8). In relation to Buchan, the caves reserve was recognized as a regional example of a natural area intensively managed for high levels of recreation. The caves in the Buchan –Mid Snowy River area were recognized for their geological values, and 'many have a history of aboriginal use as well' (NRE & PV, 1997: 19).

The ecotourism strategy noted that in recent years interest in experiencing Aboriginal art, history and culture had increased (NRE & PV, 1997: 21). In terms of visitor amenities and services, opportunities were found to exist for on-site information boards, interpretive walks, and guided tours and activities. Along existing tourist drives and key walking tracks more on-site interpretation facilities were recommended. Aboriginal culture was identified as one of the more popular themes to be considered (NRE & PV, 1997: 36). Guided tours based on Aboriginal culture were seen as an opportunity. A special event based around the Bataluk Cultural Trail was another possibility. The Strategy encouraged Aboriginal communities to explore the benefits they may derive from further involvement with cultural tourism (NRE & PV, 1997: 44).

3.7 Conclusion

Although the presence of caves in the Buchan district was first mentioned in 1840, the first known visit from staff from the Surveyor-General's department did not take place until 1854. Nascent tourism began to occur from the 1870s. In 1889 the first systematic geological survey was undertaken and recommendations made to develop the site into a tourist attraction similar to the Jenolan Caves in New South Wales replete with a caretaker and the installation of electricity. Further recommendations by a geologist with the Mines Department in 1900 saw the reservation of some 65 hectares of Crown Land near Buchan by the Department of Crown Lands and Survey in 1901, and an additional 48 hectares in 1902. The camping reserve at Buchan Caves was proclaimed in 1930 and in 1938 the Buchan Caves National Park was officially opened (see Fig.3.4).

The two show caves at Buchan, Fairy Cave and Royal Cave, were 'discovered' in 1907 and 1910 respectively. The opening to the Fairy Cave was enlarged using some gelignite and after the construction of pathways and wire netting to protect the stalagmites and stalactites, cave was opened to the public in late 1907. Royal Cave was opened in late 1913 after reserve employees cut through a solid block of marble and used a large quantity of explosives to blast through 150 feet of rock. In 1920 a generating plant was installed at the complex which remained in use until 1969 when the site was connected to the state electricity grid. A notable feature of the Buchan Caves reserve is that it has been modelled on the United States of America National Parks Service, especially its adoption of 'parkitecture' styles for some of its buildings.

In terms of the agencies responsible for the protection and development of tourism at the site, it is possible to identify a mixture of government departments and geologists, the local Buchan community, regional progress association, and key inviduals such as Frank Moon.

The Buchan Caves have significant Aboriginal values, especially the widespread heritage of caves as places where the Nargun and wicked and mischevious Nyols lived. For this reason Aboriginal people were not in the habit of venturing deep into the limestone caverns.

References

Aitken, R. (1994). *Classification Report – Buchan Caves Reserve*. Melbourne: National Trust of Australia (Victoria).

Austin, P. (1992). *A Dictionary of Gamilaraay Northern New South Wales*. Bundoora: Department of Linguistics, La Trobe University.

Australasian Sketcher with Pen and Pencil 28/8/1880.

Bairnsdale Advertiser and Tambo and Omeo Chronicle 2/5/1889; 10/8/1897; 18/1/1902; 4/3/1902; 15/4/1902; 19/4/1902; 29/4/1902; 26/8/1905; 22/2/1906; 26/4/1906; 6/12/1906; 19/3/1907; 23/3/1907; 4/4/1907; 23/12/1911.

Bendigo Advertiser 1/5/1889.

Billis, R.V. & Kenyon, A.S. (1974). *Pastoral Pioneers of Port Phillip*. Melbourne: Stockland Press.

Boadle, P. (1991). *The Management of Karst and Cave Resources in the Buchan and Murrindal Area*. Draft prepared for the Bairnsdale Region Department of Conservation and Environment, September.

Browne, R.S. (aka Tanjil). (1886). *Our Guide to the Gippsland Lakes and rivers ...* Melbourne: M.L. Hutchinson.

Buchan Sesquicentenary Committee. (1989). *Bukan-Mungie 150 years of Settlement in the Buchan District – 1839-1989*. Buchan: Buchan Sesquicentenary Committee.

Bulmer, H.D. (n.d.) *Beautiful East Gippsland containing views of Bairnsdale along the Prince's Highway The Lakes and Buchan Caves*. Bairnsdale: Howard D. Bulmer.

Calnin, D. (1997). *National Estates Grant $6000 Interpreting Aboriginal Significance – Buchan*. A Report to National Estates Grant Program, 2 pp.

Clark, I.D. (1998a). *Place Names and Land Tenure – Windows into Aboriginal Landscapes: Essays in Victorian Aboriginal History*. Melbourne: Heritage Matters.

Clark, I.D. (ed.) (1998b). *The Journals of George Augustus Robinson, Chief Protector, Port Phillip Aboriginal Protectorate, Volume 4: 1 January 1844 – 24 October 1845*. Melbourne: Heritage Matters.

Department of Lands and Survey. (c.1909). *Victoria's Tourist Resorts. The Buchan Caves*. Melbourne: Department of Lands and Survey.

Department of Crown Lands and Survey. (1975). Buchan Caves Attract the Tourist. *Landmark The Journal of the Department of Crown Lands and Survey*, 1:10-12.

Dixon, R.M.W. & Ramson, W.S. & Thomas, M. (1992). *Australian Aboriginal Words in English – Their Origin and Meaning*. South Melbourne: Oxford University Press.

Douglas, J.G. (n.d.). *Buchan Caves: A Geological Discussion, Revised from original of JA Talent*. Melbourne: Department of Minerals and Energy, Victoria.

East Gippsland Institute of TAFE. (1996). *Bataluk Cultural Trail Cultural Mapping Project*, Final Report for The Council for Aboriginal Reconciliation, Bairnsdale, January.

Fison, L. & Howitt, A.W. (1880). *Kamilaroi and Kurnai*. Melbourne: George Robertson.

Gardner, P.D. (1992). *Names of East Gippsland; their origins, meanings and history*. Ensay: Ngaruk Press.

Gippsland Times 25/2/1873; 5/7/1873; 15/4/1878; 29/5/1885; 25/2/1889; 12/3/1890; 15/10/1906; 8/4/1907; 11/4/1907; 10/1/1946.

Heath, F.V. (1907). *Views of the Buchan Caves and Pyramids*. Melbourne: TC Lothian.

Henderson, K. & de Quadros, M. (1993). *The Buchan & Murrindal Caves East Gippsland, Victoria.* Williamstown: Henderson de Quadros Publications.
Howitt, A.W. (1904). *The Native Tribes of South East Australia.* Melbourne: Macmillan and Company.
Lavelle, R.H. (1952?). Buchan Caves Victoria Australia: Australia's Most Wonderful Caves. Unpublished draft manuscript in Parks Victoria archived file Buchan Caves File No. 41 (see below).
MacCannell, D. (1976). *The Tourist: A New Theory of the Leisure Class.* New York: Schoken
Massola, A. (1968). *Bunjils Cave: myths, legends and superstitions of the Aborigines of South-East Australia.* Melbourne: Lansdowne.
Minister of Lands. (c. 1925). *Buchan Caves.* Melbourne: Published by authority of the Hon. The Minister of Lands for issue to tourists, Government Printer.
Moon, R. (1985). Buchan Caves, more than just holes. *The Age*, 8/2/1985.
Morgan, P. (1997). *The Settling of Gippsland A Regional History.* Traralgon: Gippsland Municipalities Association.
North Melbourne Advertiser 19/9/1890.
NRE & Parks Victoria. (1997). *Far East Gippsland Ecotourism Strategy.* Orbost: Department of Natural Resources and Environment/Parks Victoria.
Pepper, P. & De Araugo, T. (1985). *What Did Happen to the Aborigines of Victoria, Volume 1: The Kurnai of Gippsland.* Melbourne: Hyland House.
Pepper, P. & De Araugo.T. (1989). *You are what you make yourself to be: the story of a Victorian Aboriginal family 1842-1980.* Melbourne: Hyland Press.
Pickersgill, J. (ed.) (1885). *Victorian Railways tourist's guide: containing accurate and full particulars of the watering places, scenery, shooting, fishing, sporting, hotel accommodation, etc. in Victoria ...* Melbourne: Sands & McDougall.
Portland Guardian 22/10/1906
Roberts, L. (1977). *A brief history of Buchan district and schools prepared for the Buchan school centenary.* Buchan: Buchan School Committee.
Ryrie, S. (1840). Journal of a tour in the Southern Mountains, Ref. No: ADD 204, Dixon Library, State Library of New South Wales [Reprinted *Gippsland Heritage Journal*, December 1991, 11:11-17].
Salierno, A. (ed.). (1987). *East Gippsland Past and Present.* Victoria: F. Amendola.
Seddon, G. (ed.). (1989). *The Ballad of Bunjil Bottle; AW Howitt's exploration of the Mitchell River by canoe in 1875.* Churchill: Centre for Gippsland Studies, Gippsland Institute of Advanced Education.
Seddon, G. (1994). *Searching for the Snowy: An Environmental History.* St Leonards: Allen & Unwin.
Smyth, R.B. (1878). *The Aborigines of Victoria, with notes relating to the habits of the natives of other parts of Australia.* 2 Vols. Melbourne: Victorian Government Printer.
Swift, D. (1951). *Buchan – Valley of Caves.* Melbourne: Buchan Caves Advisory Committee, Department of Lands and Survey. [Reprint from *Mining and Geological Journal*, 1:4, September 1951].
The Argus 17/11/1854; 22/5/1886; 10/4/1902; 3/10/1906; 11/10/1906; 25/10/1906; 5/4/1907; 27/5/1907; 29/11/1907; 19/11/1910; 17/3/1914; 21/3/1914; 24/3/1914; 11/10/1918; 18/1/1924; 5/2/1926; 25/7/1931; 1/8/1931; 18/6/1935; 24/11/1937; 2/12/1938; 13/5/1939; 16/11/1951.
The Courier-Mail 20/2/1936.
The Daily News 21/2/1922.
The Malvern Standard 26/4/1919.
The Northern Miner 21/4/1925.
Urban Spatial and Economic Consultants Pty Ltd. (1996). *Gippsland Tourism Infrastructure Audit.* October. West Melbourne: Urban Spatial and Economic Consultants Pty Ltd.

Urban Spatial and Economic Consultants Pty Ltd. (1997). *Lakes and Wilderness Regional Tourism Development Plan*. September. West Melbourne: Urban Spatial and Economic Consultants Pty Ltd.

Wells, J. (1986). *Gippsland: a place, a people and their past*. Drouin: Landmark Press.

Whitcombe, F. (c. 1907). *Guide to Buchan Caves and Gippsland Lakes, illustrated by HD Bulmer and NJ Caire*. Cunninghame: Cunninghame Progressive Association.

Government Departmental Files

Parks Victoria Files: Buchan Office

Buchan Caves Reserve: Publicity', Department of Lands and Survey, Archived file, Buchan Caves File No. 41.

'Caves and Tourism', Department of Lands and Survey, Archived file, no file number.

Appendix 3.1 Intepretive Signage Installed in 1996

People in the Ice Age

– Who lived here?

Aboriginal people from the Ganai (or Kurnai) and Bidiwal nations lived in the Buchan area. They moved around in small bands of several families, and set up base camps from time to time to exploit the local resources.The open grasslands of the Ice Age made good hunting grounds. But people also needed shelter, access to water and corridors for travel. This made the few caves and rock shelters close to streams ideal places to live.

– Where and how did they live?

One of the most significant Ice Age human occupation sites in Victoria is at Clogg's Cave on the Buchan River. Starting around 18,000 years ago, people lit campfires inside the cave for warmth and cooking food such as possums, wallabies, kangaroos and koalas. The cleaned animal skins were rubbed with burnishing stones until soft, cut into shape using sharp quartz flakes, and then sewn together using bone awls and sinews from kangaroo tails to make cloaks. But when the Ice Age finished some 13,000 years ago, Clogg's Cave was abandoned until 1,000 years ago when local Aboriginal people occasionally used it during hunting expeditions. This long history makes Clogg's cave a very important place for present-day Aboriginal people.

– Aboriginal people today

The local aboriginal communities invite you to find out more about their rich history, heritage, and culture at some of the region's aboriginal cultural centres:

- Ramahyuck Aboriginal Corporation, Sale
- Krowathunkoolong (the Keeping Place), Bairnsdale
- Moogji Aboriginal Council, Orbost
- Nulluak Gungji, Cann River

The Bataluk Cultural Trail of East Gippsland also provides a fascinating discovery of local Aboriginal heritage. Enquiries can be made at any of the above centres.

Funded by Australian Heritage Commission National Estate Grant Program

The Font of the Gods, Royal Cave, Buchan. Showing the Twelve Apostles.

Fig.3.1: The Font of the Gods, Royal Cave, Buchan. Showing the Twelve Apostles (Source: Bulmer, N.D.)

Giant white stalactites and stalagmites in the Fairy Cave, showing Mr. Frank Moon, the original discoverer of the Buchan Caves.

Fig. 3.2: Giant white stalactites and stalagmites in the Fairy Cave, showing Mr. Frank Moon, the original discoverer of the Buchan Caves. Source: Bulmer, N.D.

Fig. 3.3: Buchan Caves, glass lantern slide by John Henry Harvey, J.H. Harvey Collection, State Library of Victoria Pictures Collection.

The slide shows wire netting installed to protect the stalactites.

Fig. 3.4: The Entrance to Buchan Caves National Park, c. 1948, Postcard: Gelatin silver photograph. State Library of Victoria Pictures Collection H95.28/6

4 Bunjils Shelter, Black Range Scenic Reserve by Ian D. Clark

This chapter presents a detailed history of the Aboriginal art site known as Bunjils Shelter situated in the Black Range Scenic Reserve, near Stawell (see Fig.4.1). It is the only known site in Victoria to contain bichrome figures and an anthropomorphic figure whose identity is known. The existence of the art site had been rumoured among the European population since the 1850s, however its existence was first confirmed when Alfred William Howitt (1904) revealed its general location, from information he gained in the summer of 1883/84 from John Connolly, a Jardwadjali speaker he interviewed at Ramahyuck Aboriginal station. Although the location of the site did not become public knowledge until 1957, its location was known to a select group of local European people from at least 1911. The incidence of graffiti also dates from this time. The site is generally regarded to be one of the most significant of the 150 or so Aboriginal art sites in Victoria, and yet its management has been characterized by nagging doubts about its authenticity (see Clark, 2005).

4.1 First Phase: Sight Sacralization and Naming 1883–1957

In terms of MacCannell's (1976) modeling, the first phase in the development of attractions is 'sight sacralization' or 'naming'. During the summer of 1883-4, A.W. Howitt was collecting information for a book on the Aborigines of southeastern Australia and he visited Ramahyuck, at Lake Wellington, in Gippsland, to gather information from its Aboriginal residents (Pepper & De Araugo, 1985). Howitt subsequently published this book in 1904. In the intervening period, Maynard Ord (1896), a local journalist, published a history of Stawell, in which he stated that rumours had circulated since the early days of the goldfields in the mid-1850s that Aboriginal paintings existed in the Black Range near Stawell. He noted that many people had sought to locate them but had been unsuccessful.

> Quite independently of the natural beauties of the miniature mountain, the Black Range basin and valleys have some local stories which are a little tinged with romance, and consequently with uncertainty, which latter quality is a parent of the former. In the early days of the goldfields settlement it was told that in some remote gorges of the Black Range there were mysterious caverns, approachable only by long labyrinth passages in which primitive drawings by an ancient and defunct peoples were existent, which were supposed to delineate the prowess of the earlier representatives of the black race long prior to the advent of European settlement and civilisation. It is still believed by some persons [in the 1890s] that such hidden caverns do exist and remain unexplored, but there seems little foundation for such a tradition. Yet the rumour circulated in the early days caused a good amount of exploration, both by boys and girls, who made sundry picnic excursions, and as these always resulted in joyousness of some kind or another, the pleasant theory has been preserved until the present day.

> It was gratifying to think that by climbing over a ridge of rocks, and then descending into a ravine, you might discover some wonder of nature or of aboriginal history that had not been known to the civilised world before. It was also invigorating for the young people to have an excuse for these wanderings, in company, under a sunny sky, by a rippling stream, in a real garden of native ferneries and other congenial surroundings.
>
> The caves with the reported drawings have never yet been discovered, but every exploration made by the most enterprising of the rising generation has revealed fresh beauties, both as regards vegetation and an unaccountable upheaval of granite, which has assumed the most grotesque and yet picturesque proportions (Ord, 1896:45).

From Ord's account it is evident that the search for the art site(s) in the Black Range was a precursor of the relationship between art site visitation and recreation. In this specific instance the fact that the location of the site was unknown added to the mystique associated with the Black Range.

In his 1904 publication Howitt divulged what he knew of Bunjil, and confirmed the existence of the art site in the Black Range near Stawell.

> All that I know of the beliefs of the Mukjarawaint is that Bunjil was once a man who was the father of all the people, and that he was good and did no harm to anyone. I may mention here as in one sense belonging to this part of my subject, that one of the Mukjarawaint said that at one time there was a figure of Bunjil and his dog painted in a small cave behind a large rock in the Black Range near Stawell, but, I [Howitt] have not seen it, nor have I heard of anyone having seen it (Howitt, 1904:491).

Howitt's notes of his conversation with this Mukjarawaint (a variant of Jardwadjali) speaker are in the Howitt Papers in the State Library of Victoria. These papers were presented to the Library in 1972 by the Howitt family. Examination of these notes has uncovered some interesting material about the site and his informant–a man known as 'John Connolly'. He is also referred to as Johnny Connolly, and at times 'Connolly' is spelt Conolly. When interviewing Connolly, Howitt made the following notes.

> Obtained the following information from Johnny Connolly, a half-caste native of the Mukjarawaint tribe. He was brought up by his maternal grandfather's brother until the age of [blank]. He subsequently was at the Lake Condah Mission. He speaks English thoroughly and with intelligence. I am a half caste my mother was never married to a blackfellow. She lived with a digger who found Pleasant Creek.

Connolly gave Howitt the following information on the art site we now know as 'Bunjils Shelter'.

> Bunjil a man, supposed to be father of all the blacks a place at Pleasant Creek at Black Range–there is a rock with a large cave under it. Bunjil is painted in it and a little dog in each side. Road from Pleasant Creek to Campbells Reef–there is a boundary riders hut about half a mile ws/w from the well. The Black Range Hut. Hut close to road follow into Pleasant Creek. After leaving the hut mountains runs to point along side road. A little up the hill from the point 60-70 yds there

is a big round stone in a sort of hollow, the mouth of the cave faces towards the hill. Bunjil does no harm, I think he does good [Howitt Papers 1053/5 (c), Ms. 9356 SLV].

Campbell's Reef is to the west of present day Moyston (Smyth, 1869; Banfield, 1974). The locations of the boundary riders hut, called the Black Range Hut, and the well are yet to be determined.

A comparison of Howitt's notes with his 1904 publication reveals that his publication does not faithfully reproduce the information he obtained from John Connolly twenty years earlier. Unfortunately the discrepancies have had a considerable impact on the history of this site. With regard to location, Connolly gave Howitt clear instructions, yet Howitt chose not to publish them. Instead he only gave the site's general location in the Black Range near Stawell. Had Howitt chosen to publish the locational data the site would probably have become public knowledge in the early 1900s, and not in 1957.

Another major variation between Howitt's notes and his publication concerns the number of motifs at the site. Howitt (1904: 491) specifically stated the site contained the figure of 'Bunjil and his dog'; however, Howitt's notes are clear that two dogs were painted beside Bunjil. Howitt's published reference to only one dog has fuelled local speculation that some of the painting was done by Europeans; certainly the belief that the second dog was of European origin has been widespread (Massola, 1957; Halls, 1967; Banfield, 1974; Clark, 2005).

The most valuable aspect of having finally identified Howitt's Bunjil informant and having dated when the information was obtained is that it finally puts to rest any lingering doubt about the authenticity and Aboriginal origin of this art site. The discrepancies between Howitt's private notes and his published work could only become public knowledge when the notes were available to the general public in 1972. However the informant and the nature of the information were not uncovered until 1991 (Clark, 1991). Fortunately as part of the author's doctorate research I had read the Howitt Papers reasonably carefully, scrutinizing them for information on traditional spatial organization. Having established the identities and tribal affiliations of Howitt's informants from western Victoria, when I began to research the history of the management of this art site I was reasonably certain I knew the name of the Mukjarawaint informant. A quick look at the Howitt Papers in April 1991 uncovered the reference to the art site and confirmed that the informant was John Connolly.

In the absence of access to Howitt's private notes, the search for the art site continued after the publication of his 1904 ethnography. However, the confusion of having two Black Ranges in relatively close proximity on the east and west side of the Grampians, meant that the search was sometimes concentrated in the wrong mountain range. An example of this confusion is found in *The Ararat Advertiser* of 11 April 1929, where A.S. Kenyon, President of the Ethnological section of the Field Naturalists Club, Melbourne, and a former resident of Ararat, wrote 'Dr. A.W. Howitt was told by a black fellow at Tyers that in the Black Range, the range west of the Glenelg, there was a cave with a painting of Baiamai, the Great Spirit. Search has been made for this without result'.

In January 1957 Aldo Massola corresponded with Ian McCann a member of the Stawell Field Naturalists Club, and told him that he had just come across another bit of information regarding the painted cave on the Black Range, and sent him the extract from Howitt (1904: 491). Clearly this correspondence suggests that McCann and Massola had written previously about this site. The site finally became public knowledge in 1957 largely through the efforts of Mrs. W.A. Collins, the Secretary of the Stawell Field Naturalist Club. After persistent enquiries, Collins ascertained that the location of the art site had been known to a select group of local residents since at least 1911. The precision of this date comes from graffiti painted on the wall of this site.

In 1911 the eleven year olds Eric Robson and Harry Stanton named this site 'Mancave'. In 1957 Massola named it 'Bunjil's Cave'. Within the literature a range of names has appeared including: Blackfellows' Cave; Bunjils Cave; 'Bunjil Scenic Reserve'; 'Bunjils Shelter'; Bunjil Aboriginal Art Site'; Bunjil's Rock', 'Bunjils Rock'; and 'Bunjils'. The art site obviously takes its name from the spiritual being that is painted there. Generally the orthography of the name of this being is given as 'Bunjil', however other variants include Bungil, Pundjil (Thomas in Bride 1983), and Bundjil (Hercus, 1986) (for more information about Bunjil see the chapter on Lal Lal Falls in this book).

The 1990 initiative by the Koorie Tourism Unit of the Victorian Tourism Commission and the five Brambuk communities to restore indigenous place names also sought to change the names of all public art sites in the Grampians National Park (Clark & Harradine, 1990). It was believed by both parties that many existing art site names were culturally inappropriate, misleading and incorrect. For example many of the sites are called caves, when in fact they are rock shelters or rock overhangs. In the case of this art site, the submission did not include the eastern Black Range in its recommendations, essentially because the site had in part an appropriate Aboriginal name. Nevertheless, applying the same standards to this art would suggest that the name 'Bunjils Cave' is inadequate, because the site is not a cave but an alcove within a shelter. For these reasons the Brambuk Aboriginal community in 1990 resolved to call this site 'Bunjils Shelter'. Occasionally 'Bunjil Shelter' is used.

4.2 Second Phase: Framing and Elevation 1957–Present

The second phase identified by MacCannell (1976) in the evolution of attractions is 'framing and elevation' which he argued results from an increase in visitation, when demand requires some form of management intervention, whereby the sight is displayed more prominently and framed off. In 1957 when this site was 'discovered' the National Museum of Victoria was responsible for administering Aboriginal sites. Aldo

Massola, as the Museum's 'Curator of Anthropology', became directly responsible for the management of the site. Massola was later replaced as Curator of Anthropology by A.L. West. As the art site is on Crown land, the Department of Crown Lands and Survey was the relevant land manager. Subsequent site works however were undertaken by the Stawell District of the Forests Commission of Victoria. The Museum continued to have responsibility for this site until the passage of the Archaeological and Aboriginal Preservation (AAP) Act in June 1972 saw the establishment of the Archaeological and Aboriginal Relics Office in February 1973. Under the Archaeological and Aboriginal Preservation Act 1972 an Honorary Warden scheme was established whereby local people were appointed in an honorary capacity to have responsibility for monitoring the condition of given archaeological sites. Elwyn Dennis, a local land holder, was made an honorary warden in late 1985.

The first formal inspection of the site was the 1957 inspection by members of the Stawell Field Naturalists Club and local residents. Collins was taken to the site by the Holmes brothers, Max and Jim, who lived a kilometre from the site, and had known of its existence since the 1930s. Collins subsequently notified Massola, who arranged an inspection. Massola was accompanied by Collins and McCann. Massola recorded and photographed the site and published the first illustrations and interpretation in the June 1957 edition of *The Victorian Naturalist* in an article titled 'Bunjil's Cave Found'. He recorded three motifs: the figure of Bunjil and two dogs.

4.2.1 Authentication: Aboriginal or European Origin?

Since this site was first reported in 1957 its authenticity has been questioned. One common view is that the motifs of Bunjil and the two dogs are 'fake', in that they were painted by European(s) and not Aborigines. Although Massola (1957) regarded the paintings to be of Aboriginal origin, he conceded that at first glance the figure of Bunjil does not appear to be genuine as it seems to be traced in white paint and is quite unlike the work of Aborigines.

In late 1976, VAS was asked by the Department of Crown Lands and Survey to make recommendations concerning the management of this art site. Coutts replied that as this site had not undergone close examination he was reluctant to offer any recommendations. He stated that tests were needed to ensure that the paintings were in fact genuine. On his part, he had some doubt of this. If subsequent examination proved the site was genuine Coutts stated he would recommend that the area be set aside as a Crown Land Reservation and be declared an Archaeological Area.[11] Thus the initial stimulus to substantiate authenticity came from a need to make recommendations concerning the management of the Crown Land. Continuing interest in the

11 See Item 539/76 in VAS File 50-2-4 (Pt.1) 90/5788-1.

site made it necessary for management purposes to resolve the question of whether the paintings were of Aboriginal or European origin.

In 1983, F.E. Nikkelson, a Stawell resident, corresponded with R.G. Gunn, a local rock art specialist, on the question of the authenticity of this site. Nikkelson claimed that the motif of Bunjil had been drawn in the early 1950s by a little girl who lived in the area. The other person often credited with having painted the three motifs of Bunjil and the dogs is a person named Mary Campbell, who Gunn (1983a) describes as a local eccentric who lived a few kilometres south of the site, who was said to have painted the design after coming across a statue of Buddha in the bush.

Between 1979 and 1981, five separate sets of analyses were carried out on pigment samples collected from this art site. Analyses included Emission spectroscopy, X-ray fluorescence (XRF), Infra-red spectroscopy (IRS), Xray diffraction (XRD), and Scanning electron microscopy (SEM). The final conclusion from this series of analyses was that the internal red and the white outlines of all three bichrome figures, i.e., Bunjil and both dogs had been painted using traditional Aboriginal ochres (kaolinite and iron-rich clay). Over painting of at least parts of the body of Bunjil and the second dog had occurred using a European whitewash and the red in the tail of the first dog has been added by Europeans using red lead paint probably at the same time as the 1911 graffiti was added. Essentially these results agreed with the most reliable information concerning European intervention at the site (McConnell, 1985).

In many respects the history of the authentication aspect of the management of this site has been embarrassing. Interpretation of the origin of the painting has been characterized by three views: a) that the paintings were Aboriginal in origin (Ord, 1986; Howitt, 1904); b) that some of the paintings have been added to and 'touched up' by Europeans (Massola, 1957; Banfield, 1974); and c) that the paintings have been entirely the work of Europeans (the belief of various local informants; Sullivan, 1979; and Coutts correspondence 1979). Sometime between 1979 and 1980 the site was struck from the VAS Site Register when the European view had become accepted within VAS. Despite the fact that in late 1981 SEM analysis had finally established the Aboriginal origin of the site, it was not restored to the Register until early 1983. The irony of all this is of course the knowledge that from 1972 when manuscript notes in the Howitt papers in the possession of the State Library of Victoria became available to the public they contained information that removes any doubt about the paintings' authenticity. This is not to infer that pigment analysis was not necessary to determine which pigments were of a commercial origin, and were painted over non-European ochres, rather the argument is that the question of authenticity need never have become a management issue, and the impact of pigment sampling on the art site could have been minimized.

4.2.2 Management Plans and Recommendations

In March 1973, C.E. Middleton, the Secretary for Lands in the Department of Crown Lands and Survey, wrote to J. McNally, the Protector of Relics, in the National Museum of Victoria, regarding this art site. Middleton noted that the 'cave' containing Aboriginal paintings known as 'Bunjil's Cave' or 'Bunjil's Shelter' was located on an area of unalienated Crown Land. 'Arising from enquiries into other matters in the vicinity, the question has been raised as to whether a suitable area including the site of the cave should be formally reserved or other appropriate action taken with regard to control and/or management of the site'. McNally responded in March 1973 confirming that 'Bunjil's Cave' was an important Aboriginal art site situated in a picnic ground, and is to some extent protected at present by a large metal grating. Given the implementation of the Aboriginal and Archaeological Act 1972 McNally requested time to have the matter examined thoroughly before making any final decisions. Apparently nothing came of the 1973 correspondence for in December 1976 Middleton wrote to Coutts providing him with copies of this correspondence, stating that he would be pleased to receive any comments Coutts may wish to make in respect of the management of the site.[12]

In July 1982, Coutts, the Director of VAS wrote to G. Hollingsworth, the Manager of Crown Lands in the Horsham office of the Crown Lands Department informing him that Gunn's study of Bunjils Cave had been completed and that it was clear from his report that this site was probably one of the most important rock art sites in Victoria. For this reason the Director believed the site should be declared an Archaeological Area and be properly fenced and protected at the earliest possible time. In 1982 the Land Conservation Council (LCC) recommended that the Bunjils Cave allotment be classified as a Scenic Reserve (LCC, 1982).

In June 1984 J. Hanrahan, a consultant to the Australian Heritage Commission nominated this site for inclusion on the Register of the National Estate. Accordingly the Statement of Significance was that this site was considered to be the most important rock art site in the state because (a) Aboriginal interpretation of the art is available; (b) only known representation of Bunjil in Victoria; (c) doubtless religious significance and growing significance to local Aboriginal community; (d) rarity or uniqueness of use of colour, form, technique, size, shape; (e) good state of preservation; (f) rarity of location, i.e., singular and isolated; (g) established tourist attraction; (h) educational potential; and (i) possible association between shelter and scarred tree.

In August 1984 the Department of Conservation, Forests and Lands released a draft management plan of management objectives and proposed works for the Reserve. The Region adopted many of the recommendations contained in Gunn's (1983b) report on this art site. The plan reported that the area around the art site,

12 See item VAS Registry File Site No. 7423:001.

the walking tracks between the site and the car park, and the car park itself were badly eroded. Of the five management objectives one was to protect and preserve the Aboriginal art form known as Bunjils Cave. Several works were proposed including the closure and revegetating of the existing car park and the construction of a new car park; landscaping around the granite tor containing the art site to stabilize soil erosion; the removal of graffiti; the regular removal of rubbish and further graffiti as and when it occurred. Other desired works included the construction of a walking track from the new car park to the art site and to nearby scenic points; the production of a pamphlet for the reserve; the closure of the access route to the top of the granite tor housing the art site; the construction of an interpretive centre; and declaration of the art site as an Archaeological Area under the AAP Act of 1972.

In January 1986, the Horsham Region of CFL proposed five works for the reserve: realignment of the scenic walking track to include a scarred tree; upgrading of the walking track from the carpark to the site to reduce erosion and provide easier access; removal of graffiti; placing information sheets on the information board; preparing an information sheet for visitors to be distributed from an information box located next to the information board.

In August 1988 the Black Range Land Care Group met and discussed their concerns about certain aspects of the management of the reserve: they requested the removal of the barbeques and picnic tables as they considered they placed undue pressure on the site; and erosion of walking track was a problem. In response the Horsham Region of CFL proposed to discuss the removal of the picnic tables and fireplaces with the Goolum Goolum Aboriginal Cooperative and the Stawell and Grampians District Tourism Association. A management plan, or at least a management statement would be prepared and walking track signs were going to be removed so that the track could be rehabilitated. A work crew was in the process of upgrading the walk track to the art site. Accordingly in August 1988 the Horsham Region of CFL produced a draft Management statement for the Black Range Scenic Reserve.

In the first week of June 1990, a workshop on management of public art sites within the Grampians National Park was auspiced by VAS and held at Halls Gap. Its specific objectives included the discussion of site-specific management problems and future management options; develop a list of priorities for work to be carried out; to place time-frames on proposed works. A workshop kit was produced which included profiles of the public sites, including Bunjils Cave. This profile discussed Location and description; Summary of site significance; Present visitor use; Locating the site; Adequacy of carpark facilities/signage; Quality of foot tracks to site; appearance of the site environment; Suitability/ effectiveness of current management devices; Condition of the art surface; Current interpretation available; Interpretive potential and Issues for consideration.[13] This workshop recommended that this site, along with

13 See item VAS Registry File Site No. 7423:001.

Ngamadjidj, Billimina, and Gulgurn Manja, be the developed as the primary focus for tourism related to art sites in the Gariwerd region. With regard to this site the following recommendations were made: the installation of a visitor book; the installation of the interim Bunjils Cave sign; as new signs and interpretive material is prepared drop all references to 'cave'; the more appropriate general term should be 'Bunjils Shelter'; consultations on the preparation for new material currently being prepared for the information board; the placing of a new clearly-visible directional sign on the approach road indicating the turn-off to the carpark; the installation of pit toilets at the carpark; DCE to plant prickly acacia or a similar plant at the back of the granite tor to prevent people from climbing onto the rock above the alcove; a DCE officer with necessary experience to inspect drainage/ erosion problems around the alcove and forward recommendations to Brambuk and VAS; regular maintenance by DCE staff must be ensured, including removal of litter and removal of any weeds within the grille, and monitoring for new graffiti; reassessment of existing drafts of the interpretive flyer and eventual production in a format similar to those for the Grampians National Park; preparation of a more detailed sign to be installed within the grille (Hall & Abrahams, 1990). In a subsequent meeting between Horsham DCE, VAS, and Brambuk staff the recommendations were agreed with, other than the installation of pit toilets.

In late December 2005- early January 2006, the Black Range Scenic Reserve was ravaged by wild fire and Bunjil Shelter and interpretive installations along the walking track were fire-affected. The shelter was closed to the public for several months as restoration works were put in place. In 2013 Parks Victoria commissioned Heritage Insight Pty Ltd, an archaeological consulting firm, to undertake new works at Bunjil Shelter, including redevelopment and upgrading of walking tracks and the protective cage (see Figure 4.2).

4.2.3 Graffiti and Defacements

The first known defacement of this site occurred on the 25th of January 1911, when two youths, circa 11 years of age, touched up the painting of Bunjil and one of the dogs, and added the following text: 'All This Wonderland H. Stanton E. Robson Mancave 25/1/11 H. Stanton'. This 1911 incidence is believed to be the only graffiti in the reserve until 1974 (Hough, 1987), although discussion with Keith Gidley (Pers. Comm. May 1991), the District Forester in the Stawell District of the Forests Commission, who oversaw the construction of the first protective fence in 1960, has alluded to vandalism/graffiti becoming a management problem before the cage was constructed. Compared with the extensive cover of graffiti at nearby Sister Rocks, where the landscape is similar, the Bunjil site has been subject to very low and episodic levels of graffiti. Despite the extensive distribution of rocks in the reserve suitable for graffiti, most of the graffiti occurs in three areas: the tors immediate to the art site; the car park; and the area between the cark park and the art site.

The ability to be able to positively date the graffiti and the identity of the offenders proved to be an asset in the ongoing attempt to determine the integrity of the art. In 1957 Massola was told that Robson had originally leased the locality which had since reverted back to the Crown. In 1911 Eric Robson lived in Stawell and Harry Stanton lived on a farm adjacent to the rock shelter. In 1981 Robson informed Gunn that at that time it was common practice for children to mark places they had walked to, 'just to show they had been there'. When questioned Robson confirmed that he and Stanton touched up the painting, but stressed that they did not significantly alter the existing design which, they believed, had been painted by a swagman passing through the area some years before. Pigment analysis confirmed that repainting had occurred. In 1981 when shown a photograph of the site Robson stated that there appeared to have been no further changes in the paintings as he recalled them in 1911 (Gunn, 1983a).

The 1911 graffiti at this site is scientifically and historically important. Early and dated graffiti allows the stability of the rock surface to be estimated and thus allows a minimum age for the art to be calculated. Additionally the partial removal of the 1911 graffiti from this site increased the difficulty in demonstrating the authenticity of the art.

In September 1980, the site suffered severe damage from vandals (Gunn, 1980). The vandals used yellow, light blue, black and white paints, and black felt pens and scratching were also employed. Most of the graffiti consisted of names, though a drawing and a racist comment were added. Unfortunately the most extensive damage was done by the Stawell Shire Council who covered some of the graffiti with black spray-paint in an attempt at obliterating it.

In late January or early February 1988, probably during the Australia Day weekend, the weldmesh grille at the shelter was cut open and the shelter was entered. A small fire was lit on the inside of the grille, and the rock surface surrounding the figure of Bunjil was covered with slurry made from what appeared to be crushed charcoal and an unknown liquid. The inner part of the Bunjil figure was in-filled.

4.2.4 Intervention Works: Visitors' Books

In 1964, in a paper entitled 'Aboriginal relics in Victoria', Massola discussed the Bunjil art site. A 'Visitor's Book' placed in the proximity of the shelter, so as to supply 'visitors' with somewhere to write their names other than on the walls or on the paintings, proved a great success. In attempting to verify and locate this information, a member of the work crew of the Stawell Forest District in the 1960s recollected that a visitor's book was installed, but he could not recall the length of time it was in place, nor did he know what became of the books. He was adamant that books were not installed in any other art sites in the district. A visitors' book was installed in January 1991.

4.2.5 Protective Measures: Grilles

Massola (1957), after having inspected the art site, recommended that it be protected by a wire enclosure and began to make the necessary arrangements to have the shelter protected. Massola (1969:107f) confirmed that '[i]n an effort to safeguard Bunjil's Cave from vandalism, the present writer arranged to have it protected by a wire-netting enclosure, as he has done for all the other shelters in the district. In all cases the wire was donated by the Cyclone Company of Australia, and it was erected by the Stawell branch of the Forest Commission'. In discussions with Keith Gidley he believed the site was fenced in 1960. He confirmed that the Stawell District of the Forests Commission was involved with the management of this site and stated that the Shire of Stawell had no involvement. In 1960 a work gang from the Stawell District erected the shelter using the galvanized piping and wire netting that Massola had personally trucked to Stawell. The Field Naturalists in Stawell were unhappy with the wire netting because it severely restricted photography of the paintings.

The 1960 grille was only in place for a couple of years before it was vandalized and replaced in circa 1962 by a much sturdier grille comprising iron bars rather than wire netting. Neither grilles allowed access to the alcove containing the motifs of Bunjil and the two dogs, nor did they enclose every motif at the site. Although it should be said that when they were constructed they covered the three principal motifs and the existence of the omitted motif was unknown to them.

In 1975 Michel Lorblanchet concluded that the paintings of Bunjil and the dogs were generally well preserved and adequately protected. He found, however, a very faint motif of a human figure, painted in red, outside the grille just above the mouth of the shelter. This motif was damaged by both exfoliation and lichen growth and he concluded that it was not possible to protect this human figure from deteriorating. In his report he stated that 'cleaning the wall and destroying the lichens must be avoided because the paintings cover some old fossilised lichens'. Lorblanchet's (1975) recommendations received little attention because of lingering doubts that the paintings were not of Aboriginal origin.

In October 1981, Gunn recommended re-gridding of the site to include the art outside the existing grille. In March 1983, VAS sought quotes for a re-fencing programme at this site and in late April Gunn oversaw the construction of the new grille (Gunn, 1983b). In May 1983 the grille was vandalized, and in June 1984 the gate of the grille was attacked on two occasions and the padlock smashed. Although entry was achieved on the second occasion no damage was done to the artwork. A new weldmesh grille was installed over the artwork on the week 29 October–2 November 1984. A photorail was also installed (R.G. Gunn, pers. comm.). On the 5th of November the new grille was found to have been severely damaged by vandals who had carried stone blocks up onto the top of the tor housing the art site and dropped them onto the roof of the grille. The damage took four men a full day to repair.

4.2.6 Protective Measures: Conservation

During conservation work conducted at this site in May 1988 Kosinova Thorn (1988a,b) took the opportunity to analyse the microclimate and hydrogeology of the alcove, tor, and the site, and recommended that given the problem of exfoliation of the granite, consolidation of sensitive surface areas was essential within the next five years. Water movement around the site and the lack of adhesion in the white pigment were also considered problematic for the longevity of the motifs at the site. They recommended immediate steps be taken to prevent sunlight falling onto the rock surface of the alcove during summer by the construction of shade cloths or louvers. With regard to drainage they recommended that corrective drip lines should be installed, together with selective grouting, to ensure that water running down the two fissures is discharged away from the alcove as far as possible. Subsurface drainage for the south and east sides of the tor could only be designed after further study of the site.

4.2.7 Graffiti Obliteration and Removal

After the site had been vandalized in September 1980, the Stawell Shire Council quickly instigated a clean-up operation but this consisted of a further spraying over the graffiti with rectangles of black paint that only compounded the problem. In subsequent discussions with staff at the Stawell Shire depot, Gunn learned that some of the damage he had observed had been done by Council staff as part of a clean-up operation. This entailed spray painting over the graffiti in the same colour to form solid black rectangles thus obliterating the names. This procedure is commonly employed when covering graffiti on buildings and constructions, however on natural rock the result was unfortunate. A small red stick figure outside the protective grille received a coat of spray paint with the result that Gunn (1980) believed it to be totally destroyed. After several experiments with various graffiti removal techniques, in May 1981, January 1986, and November 1986, conservators were employed in May 1988 to remove the graffiti at Bunjils shelter. They removed the 1911 graffiti, along with the spray can applications made by the Shire of Stawell in 1980.

4.2.8 Installation of Interpretive Material *in situ*

In October 1976 Coutts wrote to the Director of the National Museum of Victoria concerning the type of information that was present in some of the rock shelters in the Grampians. He considered this interpretive material to be a carry-over from the days when Massola was held in high respect and his comments were adopted by the Forests Commission and displayed in the rock shelters. Coutts had discussed these interpretive signs with Alan West and they were both of the opinion that the signs should be withdrawn and replaced by more informative and accurate descriptions. Although Coutts agreed this was the responsibility of the Survey they did not claim expertise in this field and he requested that West be approached to design the contents of the new signs. Coutts included the text the Survey had drafted for Glenisla and he requested that the Museum adopt a similar format for the sake of uniformity. The Survey would be responsible for manufacturing the signs, their cost and installation.[14]

In June 1990 Nicholas Hall made the observation that the lack of interpretive material at rock art sites in the Grampians National Park was poor. Drafts of interpretive material had been prepared since 1983, and the delay in their preparation and installation was unacceptable. He considered the new sign to be placed *in situ*, produced by VAS, had less information than the one it was replacing, and it was visually inferior. He recommended that a new detailed sign be developed for placement within the cage suitable for the site's significance. In November a temporary sign was placed *in situ* that read as follows:

> Interpretive signs for this site are being updated. If you wish to obtain information about the site, its art of Aboriginal culture in general, please call at Brambuk in Halls Gap. Sorry for any inconvenience.

In December 1990 Brambuk personnel placed a new sign *in situ*. Within three days or so it had been removed by an officer from the Stawell DCE. Apparently local residents had complained to the Stawell office about the wording of the sign which used some Aboriginal place names that were part of a proposed restoration before the Place Names Committee of Victoria. The view was taken that the sign pre-empted a decision by this committee, and accordingly the C&E officer took the action of removing the sign. After discussions between DCE and Brambuk the sign was reinstated.

4.2.9 Provision of Walking Tracks, Car Park et al

In June 1984 four people employed by the Horsham Region of CFL through the Community Employment Programme conducted work at the site upgrading the walking track to the art site, re-siting the car park and erosion control in the area as recommended by Gunn (1983b). During the week of 29 October–2 November 1984 rehabilitation works undertaken at the Reserve included the construction of a new car park, picnic ground and walking track. In May 1985 the Horsham Region of CFL through a team of Community Employment Programme workers conducted further works at the site including completing the information shelter, finished the protected fence around the art, walking tracks, a new car park, and tables and barbecue.

---

14 See VAS File 64/4/6.

Hall & Abrahams (1990) assessed that the carpark at this site was well arranged with picnic tables, barbecues and rubbish bins. However in the absence of toilet facilities people have often been observed using bushes and granite tors for this purpose and this was deemed to be a management problem. During the 1987 Gale and Gillen visitor survey, 14 per cent of those interviewed requested the installation of toilets. With regard to the walking track from the car park to the art site Hall considered the 150m track was well constructed, although he believed an indication of the distance to the site was needed at the car park. The nature trail that continues around the Reserve and includes the scarred tree, both of which are advertised in the information shelter, was unclear and visitors often had difficulty locating it.

4.2.10 Interpretation

In publications:
In his article outlining the 'discovery' of 'Bunjils Cave' Massola (1957) clearly believed the art was of Aboriginal origin. He identified two distinct phases of painting, a first phase, in red, of Bunjil and one dog, and a second and later phase which included the outlining of the original figures and the addition of a second dog. He interpreted the jacket-like body decorations found on the figure of Bunjil as being imitative of the tunics worn by soldiers in the 19th century. He considered the site the most important art site in Victoria because the identity of the figure was known. He also noted that W. Holmes who owned the land adjoining it for 34 years, had told him that the cave had always been known to his family and the painting had remained unchanged in that time.

In 1967 Halls published an article in which he attempted to explain the bichrome art style found at this site as the outcome of two distinct tribal groups having had occupation of the region. The red basal pigment he described as Buandig markings. The overlying white pigment he attributed to the Jardwadjali. Although Massola (1957) identified two distinct painting phases he made no attempt to place the art in the context of traditional Aboriginal land tenure or spatial organization. The belief that Buandig clan organization extended into the Grampian region has been widespread in the general literature concerned with the Aboriginal history of western Victoria, although there is little evidence for the claim (see Clark, 1990).

Information sheets and Tourist guides:
Because this site is generally considered a significant tourist attraction, organizations concerned with cultural tourism have sought to produce information sheets on this art site. The first information sheet was produced in 1975 by the Town of Stawell in conjunction with the Stawell and Grampians Tourist and Promotion Advisory Council. They produced a foolscap information sheet titled 'Bunjil's Shelter In The

Black Range Near Stawell'. The sheet gave a history of the site's discovery by Europeans and presented some data on Bunjil. Its contents were derived from Dawson (1881), Ord (1896), Howitt (1904), and Massola (1957).

A second pamphlet for the site was drafted in September 1984 by the Community Education and Information Branch of Conservation Forests and Lands (CFL). In November 1984, Gunn drafted text for a pamphlet on this site for VAS.[15] In November 1985, VAS circulated two mock-up pamphlets on 'Bunjils Cave'. The drafts were produced by CFL. The pamphlet was intended to be distributed both at the site and also in the information shelter.[16] With regard to the draft of the pamphlet, CFL was keen to trial it as a park note during the May school holidays. In March 1990 CFL produced a draft brochure on rock art in the Grampians National Park. Bunjils was one of the sites that were promoted.[17]

Hall & Abrahams (1990) noted that the pamphlet drafted for this site in 1984 had still not been produced. Hall believed that there is considerable potential to interpret this site in its physical setting, in relation to the traditional use of the surrounding plains. Discussion of other site types, such as the greenstone quarries, mounds and the nearby scarred tree should be incorporated. The significance of this interpretive picture is increased as it serves as a contrast to those sites in the adjacent sandstone ranges. The on-site sign needs to be designed to complement, not duplicate, the information provided in the shelter. Hall recommended that there was probably a need for a single page flyer similar to the northern and southern sites sheets produced for the Grampians National Park. These were produced by the Brambuk Aboriginal centre in 1990.

4.3 Third Phase: Enshrinement 1957–Present

MacCannell (1976) has identified 'enshrinement' as the third phase in the evolution of attractions. He understands this as having occurred when the 'framing material that is used' to protect and enhance the object of interest or the nucleus of the attraction becomes a named entity and becomes a tourist attraction in its own right. Thus in the case of an Aboriginal art site, enshrinement may be deemed to have occurred when tourists find the pathway to the site and the various interpretive signs that are placed en route and the management interventions that protect the art (drip lines; protective grilles) of interest in their own right.

15 See VAS Site Registry File 7423:001.

16 See Item 230/85 in VAS File 50-2-4 (Pt.1) 90/5788-1.

17 See VAS File 64/4/6.

Proposals for an interpretation centre:
Proposals for an interpretive centre at this site were first raised by Lorblanchet (1975) and repeated by Gunn (1983b). In August 1984 the Department of Conservation and Forests and Lands drafted a plan titled 'Aboriginal Cave Art Interpretive Structure Bunjils Cave'. A small model of the structure was constructed.[18] In November the Acting Director of the Historic Places Branch of CFL confirmed in a letter to the Regional Land Management Officer in the Horsham Region, that the Interpretive Centre as per August design was not to be constructed as it was inappropriate for the proposed function and site.

Information boards:
In September 1984, the Horsham Region of CF&L approached Jane Lennon, the Acting Director of the Historic Places Branch of CF&L for approval of a draft design of the proposed interpretive centre for the Bunjils Cave Scenic Reserve. Lennon confirmed that the proposed interpretive centre was not proceeding. In its place was proposed a simple information board costing less than $1000. Lennon requested a meeting with the new Regional Management team to discuss what designs and information were appropriate for this site and for other sites in the Grampians. At the Acting Director's request, staff in the Community education and Information Branch were requested to prepare a draft brochure and interpretive sign. This would be discussed at the meeting and sent to VAS and representatives of the Goolum-Goolum Aboriginal Co-operative in Horsham.[19]

In 1985 an information shelter was constructed but apparently the accompanying information was not installed until 1987. In June 1986, VAS received a draft copy of the text for the information board. In November 1986 VAS returned this text, presumably with comments, to the Horsham Region of CFL.[20] In December 1986 VAS received a final draft copy of the information for the interpretation board at the site from the Manager of the Horsham Region of CFL. Ben Gunn commented on the Billboard display on 27 April 1987.[21] A further draft of the contents of the Bunjils Information Board was drafted by the Horsham Region of C&E in July/ August 1990. In June 1990 Hall recommended the design of new interpretational material for the information shelter must be undertaken in consultation with Brambuk and VAS to make the most of the site's traditional and contemporary context.

---

18 See VAS File 50-2-4 (Pt. 1) 90/5788-1.
19 See VAS File 50-2-4 (Pt. 1).
20 See 737/86 in VAS File 50-2-4 (Pt.1) 90/5788-1.
21 See VAS File 50-2-4 (Pt.1).

Site visitation surveys:
From July 1986 until May 1987, the Horsham Region of CFL operated and monitored a traffic counter at this and seven other sites in the National Park. In April 1987, tourist visitation to this art site was surveyed as part of a 'Grampians Art Sites–Visitor Use Study' which aimed to provide quantitative data on visitor use of publicly promoted art sites in the Grampians; information on visitor expectations; information relevant to long term management planning in areas such as physical protection of the site, visitor facilities, educational needs, graffiti removal, and development of the visitor centres in Halls Gap. The survey by Gale and Gillen (1987a, b) revealed that over the six day study period, this art site experienced the highest number of visitations of all the art sites surveyed, averaging 34 groups or 147 individuals per day. In their report, Gale and Gillen included a draft of an information sheet on this art site. With regard to this site they made three recommendations:–that pathways leading from the art shelter at Bunjils to the scarred tree be upgraded, clearly delineated and well sign-posted or closed off so as to discourage exploration by visitors searching for more art work;–a signpost at the beginning of the walk informing tourists of the length of the walk (in time) and the nature of the walk (not a strenuous walk);–the possibility of installing toilets be investigated at the request of respondents of the survey. It was observed that a large number of visitors used the bushes and rocks at the site for this purpose, which given the large number of visitors to the site led to conspicuous fouling of the site of the peak visitor period of Easter.

4.4 Fourth Phase: Duplication

MacCannell's (1976) fourth phase in the development of attractions is that of 'duplication', when copies of the nucleus of the attraction, in this case the art site, are made available through media such as paintings, photographs, and postcards. A replica of its art was put on display in a tourist attraction in nearby Stawell in 1975, and it was chosen by Australia Post in 1984 to represent Victorian rock art in a series of eight stamps entitled 'The First Australians'.

In January 1975 J. McNally, the Protector of Relics and Director of NMV, considered there were strong moral and philosophical arguments against restoring the paintings. However given that the rock art has a finite life, it was suggested that facsimiles should be produced in close proximity to their present locations. It was considered that the scheme needed funding from the National Estate Commission. Coutts called this the 'Facsimile Project'. In April 1975, W. Cleghorn, the Director of Stawell's Mini World, corresponded with McNally on the issue of the proposed 'Stawell's Mini World Tourist Project'. Copies of the correspondence were sent to the Archaeological and Aboriginal Relics Office. The Aboriginal exhibit planned for this project was designed to be representative of the 'former Aboriginal occupation' in the Stawell district. It was felt that the most appropriate form this could take was that of a replica of Bunjils

Cave. The rationale for this was the proximity of the art site to the Town of Stawell i.e. its local significance), and the value a replica would offer the tourist industry and those concerned to minimize the impact of visitation on the site. It was regarded that a replica located in Stawell would satisfy the merely curious visitor' and 'take the heat off the art site for all but the really interested and concerned who would be more likely to respect its integrity should they seek it out'.[22] In August 1975 N. Glasby, from Glasby Bourne consultants in the World of Miniature preservation developments, submitted an application to the sites of significance committee of AIAS for pilot research funds into suitable methods of replication of rock art: they had had discussions with the photogrammetry division of the School of Surveying (Melbourne University) and it was their considered opinion that photogrammetry would accurately locate the paintings in relation to the rock contours for later reproduction and enable checking the accuracy of the facsimile, and produce a line drawing of the art for posterity. They believed this technique was not a practical method of reproducing full scale replicas of rock surfaces as it merely produced a line contour drawing of the subject, and it was not pursued further at this stage.

In 1968 Massola published *Bunjils Cave: Myths, Legends and Superstitions of the Aborigines of South-East Australia*. Although this art site was incorporated in the book's title, and a colour photograph of the painting features on the jacket cover–which incidentally has been reversed–discussion of the site is limited to a passing reference in the postscript (see Fig.4.3). In this reference Massola suggests it is possible that ceremonies were performed in front of the shelter. A photograph of the site is correctly presented in Massola (1971).

In early 1984 Australia Post asked Gunn to submit photographs of Victorian art sites. This site was subsequently selected to represent Victorian rock art in an issue of eight stamps featuring Aboriginal rock paintings entitled 'The First Australians' (see Fig.4.4). The issue was the first in a bicentennial series leading up to the bicentenary in 1988.[23] Nucolorvue Productions (n.d.) has released a postcard of Bunjil's Shelter.

4.5 Fifth Phase: Social Reproduction

For MacCannell, the final stage of sight sacralisation is social reproduction which occurs when groups, cities, and regions begin to name themselves after famous attractions. Other than the access road which has been named 'Bunjils Cave Rd', it has not been possible to find examples of social reproduction of this attraction. The local Brambuk Aboriginal cultural centre offers a guided tour which they have named 'Bunjil's Creation Tour'. In 2002, a now defunct Aboriginal tour company, 'Aboriginal

22 See VAS File 50-2-4 (Pt. 1) 90/5788-1.
23 See VAS File 6-4-17, 90/4415-1.

Dreamtime Trails', based in Ballarat, offered a tour they named 'The Bunjil Creation Tour' (see Clark, 2002).

4.6 Conclusion

Bunjils Shelter did not become a tourist attraction until it was 'rediscovered' in 1957. Given its importance and fragility as an Aboriginal art site it has progressed through MacCannell's attraction development phases within a relatively short time period. The framing and elevation phase at Bunjils Shelter occurred before there was a significant increase in visitation; and rather than have tourist demand force management intervention, it was the fragility and rarity of the site that demanded that relevant authorities protect the art, or what Gunn understands as the 'nucleus' of the attraction. The management interventions taken to protect the art such as drip lines and protective grilles equate with Gunn's 'inviolate belt' zone. The enshrinement phase at Bunjils Shelter is not a significant phenomenon, though it may be argued that this occurs when tourists find the pathway to the site and the various interpretive signs that are placed *en route* of interest in their own right. These interventions equate with Gunn's 'zone of closure'. The duplication phase has occurred in that a facsimile of the shelter was produced and displayed in a nearby theme park attraction in Stawell in 1975 and the art featured in an Australian Post postage stamp series in 1984 dedicated to the First Australians. Social reproduction of the attraction has been minimal probably due to the fact that it is not deemed appropriate for entities to name themselves after an Aboriginal art site, especially one that features a representation of the creator spirit Bunjil.

References

Ararat Advertiser. (11/4/1929). Letter to the Editor from A.S. Kenyon, President Ethnological section of Field Naturalists Club, Melbourne.

Banfield, L. (1974). *Green Pastures and Gold A History of Ararat.* Melbourne: Mullaya.

Bride, T.F. (1983). *Letters from Victorian Pioneers.* South Yarra: Lloyd O'Neil for Currey O'Neil. [Previously published Melbourne: Heinemann, 1969; Melbourne: Government Printer, 1898].

Butler, R.W. (1980). The Concept of a Tourist Area Cycle of Evolution: Implications for Management of Resources. *Canadian Geographer*, 24:5-12

Clark, I.D. (1990). *Aboriginal Languages and Clans: An Historical Atlas of Western and Central Victoria, 1800-1900.* Clayton: Monash Publications in Geography, No. 37, Monash University.

Clark, I.D. (1991). *Bunjils Shelter VAS Site 7423/1: A History of Management and Intervention.* A report to Victoria Archaeological Survey.

Clark, I.D. (2002). The ebb and flow of tourism at Lal Lal Falls, Victoria: a tourism history of a sacred Aboriginal site. *Australian Aboriginal Studies*, 2: 45-53.

Clark, I.D. (2005). An Analysis Of Challenges To The Authenticity Of Rock Art Sites In The Grampians-Gariwerd Region Of Victoria, Australia. *Rock Art Research*, 22 (2): 141-145.

Clark I.D. & Harradine, L. (1990). *The restoration of Jardwadjali and Djab wurrung names for Rock Art Sites and Landscape Features in and around the Grampians National Park*, A submission to the Victorian Place Names Committee.
Dawson, J. (1881). *Australian Aborigines*. Melbourne: George Robertson.
Department of Conservation, Forests and Lands. (1984). *Bunjils Cave Scenic Reserve–Management Plan*, Horsham Region.
Gale, F. & Gillen, J. (1987a). Visitor Survey of Aboriginal art sites within and adjacent to Grampians National Park, preliminary report. In VAS File 66-1-8 [90/5309-1]
Gale, F. & Gillen, J. (1987b). Visitor Survey of Aboriginal art sites within and adjacent to Grampians National Park, second preliminary report. In VAS File 66-1-8 [90/5309-1]
Gunn, C.A. (1994). *Tourism Planning Basics, Concepts, Cases*. Washington: Taylor and Francis.
Gunn, R.G. (1980). Analysis Requirements of Pigment samples from Rock Art Sites in Victoria, Unpublished report to VAS. Location: VAS File 8-3-1 (90/4446-1); 50-2-45 (Pt.3) [90/5838-1]
Gunn, R.G. (1983a). Victorian Rock Art Survey Quarterly Report I April-30 June 1983, Unpublished report to VAS. Location: VAS File 50-2-45 [90/5837-1]
Gunn, R.G. (1983b). *Bunjils Cave Aboriginal Rock Art Site (Site 74231 I001)*, Victoria Archaeological Survey Occasional Reports Series, No. 13.
Hall, N. & Abrahams, F. (1990).Grampians Rock Art Management Workshop, Halls Gap, June 5-8 1990, Summary and Recommendations, A Report to VAS.
Halls, F. (1967). Temples of the Dreamtime. *Walk*, 18:4-13.
Hercus, L.A. (1986). *Victorian Languages: a late survey*. Canberra: Department of Linguistics, Research School of Pacific Studies, Australian National University.
Hough, D. (1987). Vehicle Counts at Aboriginal Art Sites in the Grampians Notes, VAS Internal Working Document, 198V1. Location: VAS Registry Report No. 125
Howitt, A.W. (1904). *The Native Tribes of South East Australia*. Melbourne: Macmillan.
Howitt, A.W. (n.d.). Notes on the Mukjarawaint collected from John Connolly in Howitt Papers, Ms.9356, Box 1053/5(c), State Library of Victoria [undated but obtained summer of 1883/84].
Kosinova Thorn. (1988a). A Report on the Recent Removal of Overpaint from Bunjils Cave Together with a General Condition Report on Five Sites, A Report to the VAS. Location: VAS File 50-2-51 [90/5844-1]
Kosinova Thorn. (1988b). Bunjils Cave A Report On The Removal of Recent Overpaint, A report to VAS. Location: VAS File 50-2-4 (90/5788-1)
Land Conservation Council, Victoria (1982) *Final Recommendations South-western Area District 2*. Melbourne: Land Conservation Council, Victoria.
Lorblanchet, M. (1975). Rock Art Erosion in the Grampian Mountains Preliminary Investigation, Unpublished report to the Australian Institute of Aboriginal Studies, Canberra. Location: VAS Registry, Report No. 122.
MacCannell, D. (1976). *The Tourist: A New Theory of the Leisure Class*. New York: Schoken.
McConnell, A. (1985). 'Summary of Analyses of Pigments From Bunjils Cave', unpublished report to V AS. Location: VAS File 8-3-1 (90/4446-1)
Massola, A. (1957). Bunjils Cave Found. *Victorian Naturalist*, 74:19-22.
Massola, A. (1964). Aboriginal Relics in Victoria. Unpublished paper, National Museum of Victoria.
Massola, A. (1968). *Bunjils Cave: Myths, Legends and Superstitions of the Aborigines of South-East Australia*. Melbourne: Landsdowne Press.
Massola, A. (1969). *Journey to Aboriginal Victoria*. Adelaide: Rigby.
Massola, A. (1970). Notes on the Aborigines of the Stawell District. *Victorian Naturalist*, 87: 4-9.
Massola, A. (1971). *The Aborigines of south-eastern Australia, as they were*. Melbourne: Heinemann.
Nucolorvue Productions. (n.d.). Post card of Bunjil's Shelter, Aboriginal rock-painting, Stawell, Victoria.
Ord, M. (1896). *Stawell Past and Present*, Stawell News and Pleasant Creek Chronicle, Stawell.

Pepper, P. & De Araugo, T. (1985). *What Did Happen to the Aborigines of Victoria, Vol. 1: The Kurnai of Gippsland*. Melbourne: Hyland House.
Smyth, R.B. (1869). *The Gold Fields and Mineral Districts of Victoria*. Melbourne: Government Printer
Sullivan, M. (1979). Spectrographic Analysis, unpublished report to VAS. Location: VAS File 8-3-1 (90-4446-1); VAS File 50-2-4 (90/5788-1).
Town of Stawell in conjunction with the Stawell & Grampians Tourist & Promotion Advisory Council. (1975). Bunjil's Shelter in the Black Range Near Stawell.

Victoria Archaeological Service Files

6-4-17 [90/4415-1] Australia Post Proposal to use Bunjils Cave in Art Series.
8-3-1 [90/4446-1] Analysis of Pigment Samples From Prehistoric Art Sites.
50-2-4 [90/5788-1] Bunjils Cave – Grampians.
50-2-45 [90/5837-1] Survey of Aboriginal Rock Art in Victoria- Consultant R.Gunn.
50-2-51 [90/5844-1] Monitoring of Microclimatic Conditions in the Grampians Rock Art Shelters.
64/4/6 [90/5148-1] Grampians Area – Site Protection.
66-1-9 [90/5309-1] Grampians Visitor Survey & Relevant Data.
7423/001 Bunjils: Interpretation.

Fig. 4.1: Bunjils Shelter: Photograph Peter J. Clark 29/12/2013

Fig. 4.2: Bunjils Shelter: photograph Peter J. Clark 29/12/2013

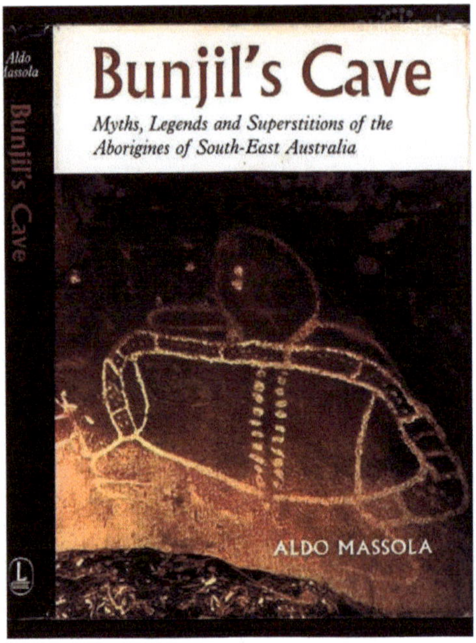

Fig. 4.3: Reproduction of Bunjil on cover of Massola's 1968 text. Note the image is reversed.

Fig. 4.4: Postcard Australian Bicentenary Aboriginal Art Bunjils Shelter 1984 (Source: Ian D. Clark postcard collection).

5 Den of Nargun, Mitchell River National Park by Sharnee Sergi and Ian D. Clark

This chapter is concerned to document the history of the development of the Den of Nargun as a tourism site utilising the theoretical constructs developed by MacCannell (1976), Butler (1980) and Gunn (1994). These perspectives provide insights into the historical maturation of a cultural or natural site into a tourism attraction. MacCannell's (1976) perspective reflects progressive development of attractions over five phases – naming, framing and elevation, enshrinement and duplication, and social reproduction. For the purpose of this study, Butler's (1980) 'tourism area life cycle model' will be correlated with MacCannell's model of the evolution of attractions in order to navigate the development and tourism history of the Den of Nargun. Furthermore, utilisation of Gunn's (1994) spatial model helps to provide an understanding of the contextual and environmental development and character of the site.

The Den of Nargun is an important geological and Aboriginal cultural site located in the Mitchell River National Park in East Gippsland and is situated approximately 50km northwest of Bairnsdale, Victoria. The evolution of the Den of Nargun as a tourism attraction is contextualised with the Mitchell River National Park's own history and will be, in part, explored in this study. The Mitchell River National Park and the Den of Nargun, as it is known today, was formally known as the Glenaladale National Park (Catrice, 1996). References to the Den of Nargun are commonly cited in older works as being a part of the Glenaladale National Park. For clarity, this study will refer to the Den of Nargun as being a part of the Mitchell River National Park and will signpost the transition from Glenaladale National Park to Mitchell River National Park. Currently, the Mitchell River National Park is approximately 12,000 hectares in size and is a part of a larger ecosystem which extends across Victoria from the Great Dividing Range and continues down to the Gippsland Lakes (Parks Victoria, 1998; Parks Victoria, 2011). The Mitchell River itself is listed as a heritage river, being Victoria's largest 'free flowing' river. The river extends down the centre of the National Park dividing the park into two sections, the east and west. The east side of the Mitchell River National Park has been widely developed for tourism enjoyment, whilst the west side remains largely untouched (Parks Victoria, 2011).

The Den of Nargun is situated along Woolshed Creek approximately 1.5 kilometres from the Mitchell River Gorge (Department of Environment and Primary Industries, 2012) (see Fig.5.1). It is set amongst warm temperate and dry temperate rainforest (Australian Government, 2011). It is believed that the cave was formed approximately 360 million years ago (Harvey, 2007). The natural occurring geological formation of the cave was shaped as a result of eroding mudstone developing beneath highly resistive sandstone. For this type of conglomerate cavern formation, the Den of Nargun is documented as the largest on record in Australia. Large stalactites are present at the entrance of the tunnel with small stalactites formed on the inner surface of the cave.

At the exterior of the cave, there is a small waterfall that runs over and down the front of the cavern entrance into a plunge pool created and serviced by Woolshed Creek (Australian Heritage Place Inventory, N.D.; Department of Environment and Primary Industries, 2012; Harvey, 2007).

Situated on the boundary between Brabralung and Brayakaulung Aboriginal country which is part of the Gunaikurnai 'nation', the Den of Nargun is of cultural significance to the local Aboriginal community (Australian Government, 2011). The Indigenous significance of caves in Victoria is noted by Clark (2007: 7) who notes that 'caves were often thought to be the abode of malevolent creatures and spirits' including the mythical creatures inhabiting Gippsland, the Nargun and the Nyol. The Den of Nargun is home to the mythical creature, Narguna who is described as being a 'fearsome creature with human arms, hands and breasts, otherwise made of stone' (Calder, 1990). Etheridge (1893: 130) has commented on this 'well known dread of caves exhibited by the Aborigines, a dread even handed down to the half-castes. Unless under very exceptional circumstances few Australian blacks can be induced to enter one of these places'. He referred to the example provided from Howitt: 'Again, our leading authority on all matters aboriginal, Mr. A.W. Howitt, in describing his exploration of the Mitchell River, in North Gippsland, mentions the 'wonderment of his sable companions, Turnmile and Bungil-Bottle, at the stalactite caverns on Deadcock Creek, regarded by them as the haunt of the mysterious creature', the 'nargun,' the 'nargun a narguna', or 'den of the nargun' (Etheridge, 1893: 130).

It is believed that the Brayakaulung tribe who lived on the western side of the Mitchell River and Brabralung tribe who inhabited the eastern side of the Mitchell River had been visiting and utilising the Den of Nargun for 5000 years (Harvey, 2007). The Den of Nargun is of particular importance to Indigenous women and it is said to be a site where women's 'initiation and learning ceremonies' occurred (Australian Government, 2011; Australian Heritage Place Inventory, N.D, para. 7; Harvey, 2007). It is also believed that Indigenous men from the local tribes will not visit the Den of Nargun as it is a sacred place for Indigenous women only (Environment Victoria, N.D.).

In terms of European land tenure, the Den of Nargun site formed part of the Glenaladale pastoral run, some 20,480 acres first taken up in October 1845 by the McLean brothers and Simon Gillies (Billis & Kenyon, 1974: 211). In July 1866 the run was subdivided into Glenaladale and Mharlooh: the former was licensed to J.H. Clough and J.W. Bogg, and the latter to Simon Gillies.

5.1 First Phase: Sight Sacralization and Naming 1875–1919

MacCannell's (1976) theory of sight sacralization identifies the first phase of attraction development as the 'naming stage'. This refers to the initial notions of naming a site and forming the final site name. Prior to the appointment of the placename, steps are usually taken to investigate attributes, features and historical poise (MacCannell,

1976). In this instance the site was specified the name 'Den of Nargun' and the nucleus is the cave itself. The aesthetic and cultural value of the nucleus and the supernatural presence of the Nargun believed to be living in the cave would motivate both environmental and cultural visitation in years to come.

It is believed that the first European to visit and record information about the 'Den of Nargun' was Bairnsdale-based goldfields warden Alfred William Howitt in January 1875 who 'found' the site during a search of the Mitchell River (Howitt, 1875). He was accompanied by two Brabralung Aboriginal guides Bunjil-Bottle aka Charley Boy and Master Turnmile aka Long Harry. Howitt published the following account of his visit to the site:

> After slowly picking my way along the rocky floor of the creek for perhaps a mile, admiring the beauty of the scene, and occasionally drinking from the rock pools of "nice cool la-en yarn" (water), as Tunmile phrased it, we came to a picturesque cavern, formed by the wearing away of the soft beds from underneath a hard course of grit. It extended in a semicircular form across the creek, the roof of the cave being the ledge over which the water falls during rainy times. The blackfellows were delighted with this "house," and planned to themselves how they could come and camp here and collect the tails of the "woorayl" (lyre bird) among the scrubs of the river, and feast on the native bears and wallabies, poor unsuspecting creatures of the neighborhood.
>
> Here we had to climb up the cliff at the side by a wallaby track to get out, but being once again in the creek bed there were the same pleasant shades as before. A little further on we came to a second cave, a wonderfully picturesque and beautiful spot. As before, a soft bed of reddish shale had been worn away by the backwash of waters falling over a hard ledge, but here the cave was higher and deeper. In front was a pool of water looking black and smooth as glass under the dense shade of the "Lilly pillys" (Acmenia). Stalactites fringed the rim of the cavern and hung in pendent rows from its roof. A huge stalactitic mass at one side joined the roof to the floor so as partly to screen the cavern, and on either hand the rocks rose up almost perpendicularly for I think not less than 400 to 500 feet.
>
> Two lyre-birds which were disporting themselves in the cavern almost delayed their departure too long. Bungil Bottle narrowly missed one with a lump of wood which he threw as quick as lightning, and Master Turnmile, being very excited, was in the act of throwing my new tomahawk at the other among the rocks when I stopped him. I preferred my new tomahawk to a dead lyre-bird, but my aboriginal seemed not to see it in the same light.
>
> While I made a slight sketch (No. 21) [see Fig.5.2] and examined the rocks, the two blackfellows looked round the cave with many wondering exclamations of "Ko-ki" at the stalactites, two of which they carried off as wonderful objects to show their friends. I was amused listening to them conversing in the mouth of the cavern. Master Turnmile, "the dandy," thought it would be a splendid place to run off to with one of the aboriginal young damsels–a house ready provided, plenty of wallabies and native bears, and a country unknown to the other blackfellows. Bungil Bottle on his part was impressed vividly by the belief that this was indeed the haunt of the mysterious creature, the "nargun," the "ngrung a narguna" (Den of the Nargun).
>
> The nargun, according to their belief, is a mysterious creature, a cave-dweller, which haunts various places of the bush. So far as I could learn, the blacks believe the nargun haunts especially the Mitchell valley, which we had just followed from Tabberabbera. What is the appear-

> ance of a nargun they cannot describe, excepting that it is like a rock (wallung), and is said to be all stone except the breast and the arms and bands. They say it inhabits caverns, into which it drags unwary passers-by. If you throw a spear or fire at it with a bullet, they say the spear or ballet will turn back on you and wound you. There is a cave in the Miocene limestones of Lake Tyers which is said to be inhabited by a nargun, with which one of the natives, "Dan's mother," according to report, had a fight. This is all I could learn.
>
> The rocks at the "nargun's cave" I found to be the usual alternations of sandstones and grits with pebble conglomerates, and here, as in the typical locality near Iguana Creek, were thick beds of soft reddish rock, apparently devoid of stratification and evidently calcareous, judging from the stalactites which depended from the roofs and sides.
>
> From the "ngrung a narguna," we found it no easy matter to escape. The ledge which formed the roof was some 30 feet above where we stood, and overhung the deep pool, except in one corner, at which a narrow ledge gave entrance to the cavern. The rocks on either hand rose up almost perpendicularly, far up above the tops of the trees in jutting sandstone ledges. The only plan seemed to be to go back down the creek and climb up the mountain sides. At last I got the two blackfellows to fall a tall straight young tree, and after much struggling we got it reared up at the corner of the roof. Bungil Bottle cut foot notches with his tomahawk and ascended as easily as going up stairs. I followed without my boots, and long before I had walked up the notches of these 30 feet of height, heartily wished, for once, that my feet had been as hard and horny in the soles as those of my black attendants.
>
> The roof of the cavern was another succession of sandstone beds, nearly horizontal, and forming as it were broad steps up the creek. But the trees were now not so high, and the sides of the glen widened out into a gully. The heat and smoke were excessive after coming from among the cool shades below (Howitt, 1975: 219-221).

In correspondence with his father, William Howitt, Alfred Howitt, noted what his Aboriginal companions said about this location (Howitt, 1971: 200):

> The blacks said it must be the home of the 'Yabbung', a mysterious creature which they believe haunts these mountains where they were living in caves and holes and preying upon the blacks when it can catch them. If you fire at it they say the bullet will turn round and wound you – or the spear thrown will turn back and pierce the thrower. The name of this cave is therefore 'Bunga Yabbunga' or the 'Yabbung's Home' ...

Mary Howitt (1971: 200) noted that Bunga Yabbunga is now called the Den of Nargun and is a popular picnic place for sightseers. Howitt's account of his visit to the Den of Nargun was published in the Geological Survey of Victoria in 1875, along with a drawing of the site (see Fig.5.2).

Calder (1990: 207, 209) discusses Howitt's journey to the Den of Nargun:

> Howitt had come down the river from Tabberabbera in bark canoes built by the two Aborigines who accompanied him. The rapids in the Mitchell proved too much for them, however, and they abandoned the canoes and continued overland, striking up Woolshed (Deadcock) creek where they soon came across the Den of Nargun. This valley forms an impressive contrast the surrounding country at any time, but it must have been particularly dramatic on that day, as bushfires were raging amid the dry country above them while they, and large numbers of frightened birds

> and animals, remained protected in the valley below. The den formed an apparently total block to further progress, but the Aborigines soon felled a tree so that it leant against the overhang of the cave. They then cut steps in the tree, and all were able to proceed. Meanwhile, Howitt busied himself with 'a slight sketch of' 'this wonderful picturesque and beautiful spot'.

> It is documented that the Aboriginal name for the site was recorded as 'Ngrung a Narguna'. Howitt recorded the site name as it was spoken to him from the Indigenous guides. They described 'Narguna' as being fierce "half flesh and half stone [being] who could kill unwary people through the strength of their embrace" (Australian Government, 2011, para. 4).

In Gippsland, caves are associated with two mythical beings; the Nargun and the Nyol. In 1875, Alfred William Howitt explored the Mitchell River by canoe accompanied by two Ganai men–Turnmile and Bunjil Bottle (Seddon, 1989). Up one creek, known as Deadcock Creek, they came to a cavern now known as "Den of Nargun" [3GP-5]. Howitt noted that his companions expressed delight upon finding this cavern, and planned to return and camp there and collect the tails of the woorayl (lyrebird) among the scrubs of the river, and feast on koalas and wallabies. A little further on, they came to a second cave, fringed by stalactites. The two Ganai men removed some stalactites to show their friends. Bunjil Bottle was convinced that this was the haunt of the mysterious creature, the Nargun, the 'Ngrung a Narguna' (Seddon, 1989: 18). Howitt's companions could not describe a Nargun, beyond that it is like a rock (wallung), and is said to be all stone except the breast and arms and hands. It inhabits caverns, into which it drags unsuspecting passers-by. Howitt knew of another cave in the Miocene limestones of Lake Tyers that was said to be inhabited by a Nargun (Seddon, 1989: 18). Massola (1962) searched for this cave and found that its description matched not the presently named "Nargun's Cave" [3NN-1] but another cave, "Cameroon's No.2" [3NN-3] (Clark, 2007: 8).

R. Brough Smyth (1878 Vol. 1: 456-7) published the following entry on 'Nrung-A-Narguna':

> A mysterious creature, Nargun–a cave-dweller–inhabits various places in the bush. He haunts especially the valley of the Mitchell in Gippsland. He has many caves; and if any blackfellow incautiously approaches one of these, that blackfellow is dragged into the cave by Nargun, and he is seen no more. If a blackfellow throws a spear at Nargun, the spear returns to the thrower and wounds him. Nargun cannot be killed by any blackfellow. There is a cave at Lake Tyers where Nargun dwells, and it is not safe for any black to go near it. Nargun would surely destroy him. A native woman once fought with Nargun at this cave, but nobody knows how the battle ended. Nargun is like a rock (Wallung), and is all of stone except the breast and the arms and the hands. No one knows exactly what he is like. Nargun is always on the lookout for blackfellows, and many have been dragged into his caves. He is a terror to the natives of Gippsland. How Bungil Bottle behaved when he came in sight of a cave at Deadcock Creek in Gippsland, and what kind of a being Nargun is, and where he dwells, and how he behaves, are well told by Mr. Alfred Howitt.

Aldo Massola (1969), an anthropologist who conducted research into Aboriginal sites in Victoria abbreviates 'Ngrung a Narguna' to 'Nargun'. Massola (1969: 191) details his conversation with the local Indigenous people of Lake Tyers who describes the Nargun as:

> Those half human and half animal beings who lurk in deep underground recesses and who occasionally call out Nga-a-a-a. They are feared exceedingly, for they eat blackfellows, and their stone bodies have the power to turn spear or bullet back on the thrower, so they cannot be killed. The legend of the Nargun, was also said to exist at Lake Tyers at a place called the Devils hole or Ngrung. However, according to the Aboriginal legend, The Nargun was "vanquished by Lilly" who was "the only person to ever beat the Nargun". Later, Lilly fell into Lake Mystery and her body was taken by a Bunyip (Massola, 1969: 191).[24]

During the immediate years following Howitt's 'discovery' of the Den of Nargun in 1875, there is little known about any tourist or visitor activity. As described in Howitt's account, given the harsh and dense terrain that needed to be trekked in order to get to the Den of Nargun, it can be suggested that there was little tourism activity during this time (Harvey, 2007). The next account of visitation is revealed by the *Bairnsdale Advertiser and Tambo and Omeo Chronicle* (26/1/1884) in an article entitled '*How I found the Nargun's cave*' by a local land selector named J.D.S. The newspaper editorialised the article, noting that 'We believe it is not generally known that there is such a cave in this district, but it is near Lake Tyers. One of the myths of the aborigines is 'Ngrung-a-Nargun". The superstition is that a monster known as "Nargun" dwells in a cave, and though it cannot be killed by the blacks, it will kill and eat any black who approaches near the cave. It is believed that if any one throws a spear at "Nargun" the spear will return and wound the thrower. The cave referred to at Lake Tyers has always been regarded with dread by the blacks'.

> HOW I FOUND NARGUN CAVE. (By J.D.S.)
>
> When first I came to my selection, now some years ago, the blackfellows used to tell me, "Nargun will eat you." Thinking it was only their way of joking I took no notice of what they said, but when I became better acquainted with them, some of them very seriously told me, Nargun, has a cave near you and will be sure to find you out some day. One of the more intelligent of them informed me that this was a monster like a woman, but much larger, very strong, and covered with scales like a fish, that she had a great taste for eating blackfellows, and if one of them came near her cave he never got away, but was sure to be devoured. I find in Mr. Brough Smyth's interesting book about the aborigines that he mentions among their myths this of Nargun, and, that this being is supposed to live in a cave near Lake Tyers. These various statements made me wishful to discover, if possible, the residence of so formidable a neighbor. Mr. Bulmer informed me that he had seen the cave, but it was so many years ago that he had forgotten the precise locality. I often tried to induce the blackfellows to show me where it was but could not succeed; they positively declined to be eaten. I was too busy with work on my selection to give time for

24 Phillip Pepper (1980: 57) recounts a story of a group of Ganai men including Big Charlie, Big Joe, and Short Harry who killed a Nargun in a cave near Tooloo. Pepper describes the Nargun as 'seven feet tall and went out at night to hunt the children and eat them'. It was also called the 'Hairy Man'.

exploring, but sometimes on a holiday would search the gullies in the direction of the mysterious cave, though hitherto without success. Last week two young men, who were rusticating in this neighborhood, one a school master, who, tired of teaching the young how to shoot, for a change came to shoot wallabies; the other a candlemaker, came to my selection to recreate. One day was set apart for trying to find Nargun cave. We made an early start, and after pushing our way through some very rough country, came to a large gully that I thought was somewhere about the place I wished to find.

Proceeding down it for about half a mile we came to where some rocks had fallen, and half hidden by ferns and bushes was the entrance to the cave. We scrambled over the rocks and found an opening high enough for us to stand in upright. The roof came lower, and we had to stoop, then crawl on hands and knees, and our progress was neither rapid nor pleasant. We soon got tired of it, and reaching our candle as far as possible, could see no change for the better, so we backed out, and were not sorry to get into daylight again. We unanimously resolved that if this is the famous Nargun cave it is not worth coming to see. I suggested that instead of returning the way we came we should keep down the gully and follow the creek towards home as the scenery was very beautiful. All were willing, so away we went, the school master leading. Very soon he let a great shout out of him, "Why, we have not been in the cave at all, here it is." And sure enough the gully had come to an abrupt end; a spur of the range had cut across it, and in the face was a large rent or crack some twenty or thirty feet high, and just wide enough for a person to walk in. It was a wild looking place, and made us think of those lines in the "Fairie Queen,"

Ere long they come where that same wicked wight
His dwelling has, low in a hollow cave,
Far underneath a craggy, cliffy pight.
Dark, doleful, dreary, like a greedy grave
That still for carrion carcases doth crave;
On top whereof aye dwelt the gastly owl,
Shrieking his baleful note, which ever drave
Far from that hunt all other cheerful fowl,
And all about it wand'ring ghosts did wail and howl,
And all about old stocks and stubs of trees.[25]

Spenser might have been here when he wrote this, for we have the cave, the owl, (the candlemaker shot it though) old stocks and stubs of trees, but where is Nargun? Our candle had got very short through exploring the first cave, so we had not much time to give to this. For about four hundred yards it continues narrow, and then opens out wide, and is much higher, perhaps forty to fifty feet high, but when we got this far our light was nearly gone, and much to our regret we had to return. A creek flows through the cave, but is dry at present. The drift wood shows that sometimes it is four or five foot deep. The mystery of the cave has yet to be explored, how far it goes, where is its outlet, where did Nargun cook the blackfellows, and other interesting particulars remain to be discovered. If some of our enterprising summer visitors solve these problems, well and good, if not, should we live to see another holiday, we may try again.

25 Edmund Spenser – The Faerie Queene, Book 1.9.

Julian Thomas, aka 'The Vagabond', paid a visit to the Lake Tyers district in April 1886, and his account of his visit reveals the identity of LDS and also adds to our knowledge of the Den of Nargun:

> Mr Stocks, however, has not long been a farmer. He is an old Ballarat man, but has taken up this selection on the eastern shore of Lake Tyers as a haven of rest. There are patches of good black soil, and fruit grows well here, but the cost of clearing is very great. The Lancashire man, however, appears happy, and it is a pleasure to me to meet him, and have a talk about Old England. I am also highly interested in the description of the cave situated a mile from hence. The head waters of this lake have now become but a narrow stream known as the N'grung Creek. The cave is situated on its banks, and runs right through the limestone cliff, being more than a quarter of a mile long. A stream of pure water flows through the cave into the creek. In places it is 30ft high, and sometimes one walks through the narrowest of passages, at others the cave spreads out into wide chambers. One thinks that this would have been a magnificent place for the retreat of any outlaws or for blackfellows to camp and make their homes in. But the natives will not go near it. They dread this cave as much as the inhabitants of Tanna do the vicinity of the volcano of Yasur.[26] The latter, however, have reason, for the hidden fires there have caused earthquakes which have wrecked the country, and the perpetual eruptions and streams of burning lava are sufficient to scare even white people. But the Gipps Land natives are frightened by a tradition of a demon who inhabited "Nargun's Cave." An evil spirit more powerful than all the m'raajti, a gigantic "thing" in female form, covered with scales, which is possessed with a strange taste in that it eats blackfellows. *Nargun* or *Naguna* is the title of this Lamia.[27] So not even the Christian natives of the mission will go near the cave. The foundation for this story may have been the presence of a large seal in the cavern, or from the mists of the past the legend may have descended of some monstrous creature of the prehistoric order of animal life which might have survived here. But why, I ask myself, do so many races of men embody their horrible monsters in a female form? I give it up (*The Argus* 3/4/1886).

Furthermore, in circa 1900, but definitely prior to 1904, the Den of Nargun was visited by two brothers known as the 'Morrison Brothers' who discovered the cave by chance whilst droving cattle. The two men were from Glenaladale approximately 20 kilometres from the Den of Nargun (Google Maps, 2013, Catrice, 1996). In 1904 the Morrison Brothers returned to the site with psychologist and educationist, Professor Stanley Porteus (Catrice, 1996). Porteus (1951: 271) discussed the Den of Nargun in his novel *Providence Ponds: a novel of early Australia*, which is based on his wife's forebears' experiences in settling in Gippsland, and which explored the psychological effect of human isolation:

> The cave of the Nargun has become one of my favoured refuges when any perplexities visit me. It is throughout the day a happy spot, just Australian bush, nothing grand or wonderful, merely a pleasant place, alive with the tinkle of bell-birds, small waterfalls, and glistening rock slides. I have no words for its beauty....

26 Mount Yasur is an active volcano on Tanna Island, Vanuatu.
27 In ancient Greek mythology, Lamia was a Libyan queen who became a child-eating daemon.

An historical photograph published by Department of Environment and Primary Industries (2010) captures a shooting party at the foot of the Den of Nargun in 1905 (see Fig.5.3). This provides some indication of visitation occurring during this era.

Exposure of the Den of Nargun increased as the years grew closer to the 1920s. This signals the transition of the tourist site into the second phase of tourism development. On 16 June 1917, *The Weekly Times* newspaper published an article entitled *Off the Beaten Track- Visit to 'Home of Evil Spirit' by "All Round"*. This article demonstrates the increased exposure of the Den of Nargun:

> Legends of the "Ngrung-a-Narguna," or home of the "Nargun," an evil spirit of the blackfellow, reached certain members of the Melbourne Walking and Touring Club, who had also read Dr Howitt's account of a visit he paid to it...This member writes:–It is over 40 years since Dr Howitt made a trip down the Mitchell River in bark canoes with two aboriginals. After two days voyaging down the upper reaches, through mountainous gorges and over rocky rapids, they came to the mouth of the Dead-cock creek, now known to the man-makers (but to no one else) as the Woolshed...
>
> A LONELY FARM HOUSE.
>
> A ten-day moon prevented the dusk from developing into dark, and so we walked along the bridle track round the hillsides and over the spurs till we came to the lonely farmer's home within a mile of our objective. The family had gone to a patriotic concert five miles away, and would not reach home till daybreak, sleepy and with tired feet, but ready for another dance the following night if they had the chance. With charactoristic bush hospitality, they placed their larder at our disposal so we boiled, our billy at their fire and cooked some saveloys for tea.
>
> An hour later we were carrying our swags, a borrowed axe, and lantern, down a precipitous spur leading to the mouth of the Dead-cock creek. A few hundred yards up stream, we came to a pool surrounded by a natural amphitheatre sixty feet across, and floored with the flat sandstone rocks of the district. Up-stream, the water pouring over 20ft. of cliff kept the pool replenished and on the left hand the rocks rose sheer to a height of 120ft or more. Still higher could be seen the eucalypts of the bush above, with an occasional "currajong" perched on the edge of the cliff. For sixty yards along the right hand side a huge cavern extended 20ft. or more into the side of the hill, with a ceiling formed of another almost level layer of the same sandstone as formed the floor. Here we camped for the night in a little pool of bone-dry sand. The only exit from this amphitheatre was by a narrow wombat track which wound its precarious way back and over the roof to the cave until it reached once more the almost flat bed of the creek higher up. More rocky pools, more luxuriant vegetation of an Eastern Gippsland type led to another pool, waterfall and cavern. The gorge was narrower here, and the cliffs on either side were not only unclimbable, but rose sheer to a skyline fringed with currajong and stringy barks, two or three hundred feet above. This cavern, screened with stalactites and stalagmites, was the Ngrong-a-Narguna. Here dwelt the Nargun, an evil spirit whose lower half had turned to stone, and who stretched out a long arm and captured the unwary picanninie passing by.

> SCENE UNCHANGED BY TIME.
>
> The sketch made by Dr Howitt, and published in a Mines Department report in 1876 [sic], shows that little change has taken place except that one stalagmite has grown about 3ft. towards the roof of the cavern, and that a kanooka on the edge of the pool has increased from apparently about nine inches to over three feet in diameter.[28] Nothing short of a ladder could have got us past the fall...

The earliest name for this site appears to be 'Bunga Yabbunga' meaning 'Yabbung's home' sourced from correspondence between Alfred William Howitt and his father (Howitt, 1971: 200); another early name is 'Ngrung a Narguna' meaning 'home of the Narguna, the stone-devil'. It is possible that Bunga Yabbunga is a mishearing of Ngrung a Narguna. 'Nargun Cave' is recorded in 1884; 'Nargun's Cave' in 1886; 'Cave of the Nargun' in 1951. Today the site is known popularly as the 'Den of Nargun'.

5.2 Second Phase: Framing and Elevation 1920–1985

MacCannell (1976) describes the second phase in the development of an attraction as 'framing and elevation'. This refers to the deliberate attempt to increase exposure of a site or product through opening the site for tourist exploration or enhancing visual appeal. During this stage, the nucleus or site is framed through both protection and enhancement. Protection refers to safe-guarding the authentic value of the site or object and enhancement occurs when the object is enshrined. The second stage following the typology life cycle presented by Butler (1980) is the 'involvement stage'. In this stage according to Butler (1980) tourism marketing plans begin to develop including the initial development plans with a foreseen increase in tourism numbers.

Historical records show that tourism to the Den of Nargun increased from the 1920s, with a large influx of visitors in the 1930s (Catrice, 1996). The increase in visitor numbers was the result of the increased exposure to the existence of the Indigenous site. An earlier visitor, Porteus who initially visited in 1904 with the Morrison Brothers returned to the site in 1923 accompanied by Robert Henderson Croll, a leading member of the Melbourne Walking Club who documented and published the Den of Nargun and its location in his guidebooks (Catrice, 1996). The walking club in 1923 was the first to call for the permanent reservation of the Mitchell River on 'account of the splendid specimens of various indigenous trees'. Wheeler (1991: 37) suggests that Croll foretold the gazetting of the Den of Nargun as Glenaladale National Park when, on wandering through its shaded gorges in 1923, he declared it 'a remarkable place – a museum of living trees of great age. It should be preserved as a national asset'.

28 Kanooka refers to Tristaniopsis laurina, or water gum – a medium native rainforest tree with a widely spreading canopy.

Croll's newspaper articles that were later published as *The Open Road in Victoria* (1928) and *Along the Track* (1930) opened the site for tourist exploration, starting a 'hiking craze' (Catrice, 1996: 2). *Along the Track* (1930) details Croll's journey to the Den of Nargun. Croll speaks of his pursuit to locate the Den of Nargun. As Croll looked for a place to sleep they were tempted by 'sand under, a close canopy of bushes', but their pursuit of 'the cave' encouraged their travelling souls (Croll, 1930: 80). In *Along the Track*, Croll (1930) provides his impressions of the week they spent at the Den of Nargun in 1923 (originally published in *The Argus*, 2/6/1923):

> A tinkling of bell-birds suggested water, and, sure enough, well shaded by overhanging growths, we found some isolated pools, from which, eventually to draw a meal of mountain trout and a fine eel. Above the second pool a rock barrier, thirty feet high, stretched from wall to wall, and apparently, forbade all further passage. It must make an attractive waterfall when the stream runs. On the right the water has undercut the cliff to a depth of about forty feet and for a length of 150 feet, forming a habitable cave, dry, and practically waterproof. In this Ngrung a Narguna (home of the stone devil) as the natives named it to Howitt, we lived for five days, and a comfortable home it provided. Blocks and stones fallen from the roof provided seats and tables, ledges of the back wall seemed made for storing provisions, and the easterly aspect of the opening allowed the glorious moon for last Easter to flood it with light every evening. ... Despite the dreams that should have come in such a place, we saw nothing of the original owner, the nargun, described as "all stone, but his breast, and his arms and his hands", who lurked in caves to devour people, and against whom all weapons were useless. But one night we woke to a dreadful scream, which drove all sleep away. Near at hand in the gorge followed a blow as though a blunt axe had struck hollow timber. 'Hough', 'hough', 'hough' went the flail-like beat until a dozen or more had been struck. Then all was silence. In the morning we saw no sign of the mysterious visitor. The sun shone, the birds sang and disported; the evil spirit had been exorcised (Croll 1930: 80-81).

In December 1926, Charles Daley led a party of ten members of the Field Naturalists Club of Victoria on a Christmas excursion to Deadcock Creek and the Mitchell Gorge. Of the Den of Nargun, Daley wrote the following:

> The towering cliffs, the beautiful, lofty trees, the striking features of the cave, the "Nargun's Den," the twisting lianas, the attractive beauty of the setting, with its strangeness and harmonious blending of color and form, compose a picture that can only be imperfectly described, but will long live in memory. It is unique, a place apart, a restful retreat, secure in its situation from storm or devastating tire affecting the hills above (Daley, 1927: 300).

In the 1930s the Field Naturalists Club of Victoria–which was founded in 1880 and continues to operate at different locations throughout Victoria–began instigating movements to have the site protected. In 1938 they had some success when the area of Woolshed Creek including the Den of Nargun was declared a sanctuary (Field Naturalists Club of Victoria, 2012; Catrice, 1996). It wasn't until the 1950-60s that interest in protecting the sight was in the forefront again.

The Bairnsdale Field and Naturalist Club appealed to the public for support in protecting the Den of Nargun through the publication of an article in the *Bairnsdale*

Advertiser in 1962 titled *An Early Account of Visit to the Den of Nargun*. The article retold Howitt's journey to the site and states that:

> Efforts are being made by the Bairnsdale Shire Council to have a large tract of country, including the Den of Nargun area in the vicinity of the junction of Dead Cock Creek and the Mitchell River, declared a National Park...The [Field and Naturalist] Club is also of the opinion that the area should be preserved as National Park... (*Bairnsdale Advertiser*, August 1962).

Following appeals as illustrated in the *Bairnsdale Advertiser* (1962), in another article 'To the Den of Nargun Majestic Cove' published in the *Gippsland Pictorial News* in December 1962, it was announced that 'This month the "dens" on Deadcock Creek and the adjacent land were proclaimed a National Park'. The article proceeded to increase exposure to the site through restating how to get there:

> Follow Dargo road to Den of Nargun sign and along turn-off to gate. Pass a cattle grid and gateway in two miles to stopping point for vehicles (near deserted house). Follow house in general left direction into scrub. Trail is overgrown but easy to follow. Mitchell Gorge appears after about a mile. Walk to river and turn upstream to creek bed. Turn left to lower shelf of rock. Up small track at bottom end of overhang, along creek bed for mile to upper den.

Shortly after the declaration of Deadcock Creek (including the Den of Nargun) as a National Park, the area including Woolshed Creek Sanctuary was renamed to form the Glenaladale National Park in 1963, following the donation of 163 adjoining hectares from Australian Paper Manufacturers Ltd. In 1963 the Den of Nargun was formally registered as part of the Glenaladale National Park (Harvey, 2007; Catrice, 1996). Naturalist Norman Wakefield (1967: 5) confirmed that the Glenaladale National Park was established by Government gazettal on 13 November 1963, and that its focal point was the geological formation known as 'Den of Nargun'. Wakefield (1967: 8) notes that 'For many years the Den of Nargun and its surroundings were privately owned, and the general area was eventually acquired by Australian Paper Manufacturers Ltd. However, that firm generously agreed to donate the 403½ acres to the Victorian Government, and thus it was possible for the place to be made a National Park'.

On March 22, 1965, a letter was sent to the National Museum of Victoria from R.S. Yeates, from the Bairnsdale Historical Society. This letter highlights increased visitation following the creation of the Glenaladale National Park. It is unknown whether anything eventuated from this letter; however it goes some way to substantiating an increase in tourism numbers as reflected by the bell-shape curve in Butler's (1980) tourism area life cycle model. Another key development occurred in 1980 when the Den of Nargun was registered as an historic site by the Australian Heritage Database (Australian Government, 2011).

In terms of Gunn's (1994) spatial model the nucleus is surrounded by the inviolate belt which can function to protect and enhance, both physically and geographically, the natural and cultural sanctity of the nucleus. The inviolate belt was essentially fabricated in 1938 with the declaration of the Woolshed Creek Sanctuary and then reaffirmed in 1963 when the Den of Nargun became protected under the Glenaladale National Park and reaffirmed in 1980 (Australian Government, 2011; Harvey, 2007; Catrice, 1996). Physical barriers to protect the geological formation of the nucleus, and thus function as an inviolate belt, were not constructed. However, the natural plunge pool and rock formations may have acted to provide some natural protection by deterring tourists from entering the cave itself.

5.3 Third Phase: Enshrinement 1986–Present

MacCannell's (1976) third phase in the evolution of attractions is known as 'enshrinement'. This takes place when the site has graduated from framing and elevation and proceeds to increased exposure and visitation. Aligned with this stage is the continuation of Butler's (1980) tourism area life cycle model including three stages 'development', 'consolidation' and 'stagnation'. Development is characterised by changes in site structures and facilities in order to increase positive tourism experience and cope with tourism demand that continues to increase due to well sought and developed marketing plans set out in the prior involvement stage. Visitor numbers will continue to increase resulting in the graduation into the stage of consolidation. Marketing progresses focussing on an increase in economy driven advertising as the local economy becomes more prominently shaped by the tourism industry.

The next stage identified by Butler (1980) that corresponds with the enshrinement period described by MacCannell (1976) is stagnation. Stagnation will be a result of the tourist attraction reaching its '[c]apacity levels for many variables will have been reached or exceeded, with attendant environmental, social, and economic problems' (Butler, 1980: 8). Butler (1980) also suggests that initial site attractiveness which initially drew people to the nucleus will supersede as a result of social and cultural changes rather than the physical attractiveness of the site. Corresponding with MacCannell's model and Butler's tourism area lifecycle model is the 'zone of closure' described by Gunn (1994). The zone of closure is the final stage presented in the spatial model and represents the area surrounding the nucleus and inviolate belt. Typically, the zone of closure refers to visitor services and facilities which promote accessibility and enhance the visitor experience (Gunn, 1994).

The enshrinement stage began in 1986 when the Den of Nargun became a part of the Mitchell River National Park. The Mitchell River National Park formed when the Glenaladale National Park and other adjoining land was merged to form one title (Catrice, 1996). This change in composition of land in which the Den of Nargun is situated increased exposure to the site. Increased exposure occurred as a result of the Mitchell River National Park offering a variety of different tourist and leisure attractions. The Mitchell River National Park offers camping, four-wheel driving and leisure opportunities in the south end of the park where the Den of Nargun is located. The

Den of Nargun was also promoted through the newly-created Bataluk Cultural Trail which was developed in order to increase exposure of Aboriginal sites and increase Indigenous tourism in Gippsland (Parks Victoria, 1998).

The Bataluk Cultural Trail website notes:

> The Nargun is a large female creature who lives in a cave behind a waterfall in the Mitchell River. The Den of Nargun is a place of great cultural significance to the Gunaikurnai people, especially the women. Traditionally Gunaikurnai men were not allowed down to the Den of Nargun or the Woolshed Creek valley. Gunaikurnai men respected this traditional law and still do today. Please treat this place with respect. Stories were told around campfires about how the Nargun would abduct children who wandered off on their own. The Nargun could not be harmed with boomerang or spears. These stories served the dual purpose of keeping children close to the campsite and ensuring that people stayed away from the sacred cave. The Den of Nargun is a special place for women and may have been used for women's initiation and learning ceremonies. The walk into the Den takes approximately 15 minutes each way, or 45 minutes as a circuit walk via the Mitchell River. Note that there are some steep sections along the walk.

The zone of closure identified by Gunn (1994) includes the existing Den of Nargun Picnic Area, Woolshed Creek Camping Grounds, Old Weir Lookout, Deadcock Bend and Billy Goat Bend Camping Grounds (Parks Victoria, 2011). Den of Nargun Picnic Area facilities included toilets, fireplace, picnic tables, shelter and lookout (Parks Victoria, 1998). It is believed that from 1986 when the Mitchell River National Park formed, the majority of tourist facilities were constructed marking the beginning of Butler's (1980) development and consolidation stage.

By 1998, visitor statistics gathered by Parks Victoria showed that Mitchell River National Park attracted approximately 56,000 tourists annually with one of the most popular sites for exploration being the Den of Nargun. The management authority, Parks Victoria, responded to these figures in order to sustain tourism numbers and promote positive tourism experiences (Parks Victoria, 1998). Major development to the Mitchell River National Park and the Den of Nargun's immediate zone of enclosure (walking track, access roads and Den of Nargun Picnic Area) began in 1998 and are identified in the Mitchell River National Park Management Plan. In 1998 the proposed park improvements relative to the Den of Nargun, included upgrading the service road between Billy Goat Bend and the Den of Nargun from open to public to Class 1 which allows full 2WD access; installing interpretive signage at Den of Nargun Picnic Area and along walking tracks to the nucleus highlighting the cultural and natural significance of the Den of Nargun; increasing car parking at Den of Nargun Picnic Area; and enhancing the existing facilities and creating recreation opportunities at the Den of Nargun to increase visitor enjoyment (Parks Victoria, 1998) (see Fig.5.4).

Since the Mitchell River National Park Management Plan was proposed and implemented in 1998, there is little evidence of any significant interventions or development at the Den of Nargun (Parks Victoria, 1998). The only exception is the removal of the lookout at the Den of Nargun somewhere between 1998 and 2007 (Harvey, 2007;

Parks Victoria, 1998). Harvey (2007) has also suggested that the interpretive signage is out dated and needs to be improved.

Nevertheless, two events have occurred since the 1998 management plan that have significance for the Den of Nargun. Firstly, the announcement of the joint management between Parks Victoria and the Gunaikurnai Land and Water Corporation for the Mitchell River National Park. Parks Victoria (2012: 1) notes:

> On Friday the 22nd of October 2010 both the Federal and State Governments formally recognised the Gunaikurnai people as the Traditional Owners of over 20 percent of public land within Gippsland and Eastern Victoria. The Victorian Government and Gunaikurnai people formally signed Victoria's first settlement agreement under the new Traditional Owner Settlement Act 2010. This agreement involved a transfer of the ten parks and reserves to the Gunaikurnai "Aboriginal Title" which will be jointly managed in conjunction with Parks Victoria. Mitchell River National Park is one of the jointly managed parks within Gippsland. This agreement recognises the fact the Gunakurnai people have always been connected to their land and are the rightful people to speak for that country. These parks and reserves are cultural landscapes which are part of our living culture.

Whilst this announcement signals a change to pre-existing management structures which promotes greater Indigenous ownership, there is no evidence to suggest that dual management has had any effect on tourism at the Den of Nargun. However, this is a significant event in the Den of Nargun's history.

The only other known major intervention since the 1998 management plan is the rejuvenation of the Bataluk Cultural Trail in 2011 (which features the Den of Nargun). The Bataluk Cultural Trail tourist information brochure and website were redesigned and new sites in South Gippsland and Cape Conran were added to the self-directed tour (ABC Melbourne, 2011; Bataluk Cultural Trail, N.D.a; Bataluk Cultural Trail, N.D.b). In the following section, it will be argued that the Den of Nargun is still in the enshrinement phase identified by MacCannell (1976) and stagnation stage described by Butler (1980).

5.4 Fourth Phase: Duplication

The next phase in the evolution of natural attractions identified by MacCannell is 'duplication' or 'mechanical reproduction'. Mechanical reproduction refers to the deliberate act of producing products and artefacts inspired by the nucleus or site. These can include things such as copies of the nucleus, photographs, paintings, and postcards. MacCannell (1976: 45) suggests that the purpose of this phase, whether deliberate or unintentional is to set 'the tourist in motion on his journey to find the true object'.

Duplication of the Den of Nargun has been evident through the production of various artworks and writings over the past 40 or so years. As tourism and awareness of the Aboriginal site increased from the 1950s, duplication of the nucleus also occurred. The Nargun, the supernatural being, has been portrayed in numerous story books including *The Slaughter of the Bulumwaal Butcher* by Bruce Pascoe, *Lingio* (1866) by Angus McLean, *and The Ghost Child* (2008) by Sonya Hartnett. Additionally, Pearson Australia has published a primary school learning resource about the Den of Nargun (Pearson, N.D.). However, the most famous storybook written about the Nargun was that published by the acclaimed Australian author Patricia Wrightson. Wrightson first published *The Nargun and the Stars* in 1973 (see Fig.5.5). The storybook was re-published in 1988 with drawings by Robert Ingpen (Wrightson, 1988).

Wrightson's (1988: 7) story begins:

> It was night when the Nargun began to leave. Deep down the below the plunging walls of the gorge it stirred uneasily. It dragged its slow weight to the mouth of its den; its long, wandering journey had began. Two hundred feet above, on the broad uplands, moon light whitened the gum-trees where eagles were building. It split into the gorge to touch the tallest heads of coachwood and nettle-tree, but it never reached the black damp rocks of the bottom. Only where water slid over a great slab of cliff at the head of the gorge a glint silver light was carried down. At the bottom the water fell, with a sound-and-echo like guitar strings, into a pool that spanned the gorge. Behind this pool – behind a bead-curtain of falling water – cut broad and low into the base of the cliff was the arch way of a cave. This was the ancient den of Nargun. Here it had lain while it learnt fly and gum-trees to blossom; while stars exploded and plants wheeled and the earth settled; while the cave opened; while dripping water hollowed a pool from rock and filled it, and drop by drop built crystal columns before the cave. And all this while the Nargun slept. In time it opened slow eyes and saw light. Little by little it dragged itself from earth and moved. There came a night when it had a voice and cried down the gorge. There came a day when, crouching in shadows, it grasped at something warm and found food to mumble on. After that it ate when it could; sometimes once in ten years, sometimes in fifty. It moved down the gorge to blink at the sun, to watch a river flow, to hunt savagely; but always it made the ponderous climb back, crushing ferns and grinding moss on its way to drag itself behind the crystal columns...

Photographs and paintings of the Den of Nargun have been appearing online for many years. For example, photographers such as Dave Callaway captured the Den of Nargun and currently sells his images online (Red Bubble, 2012). Local Indigenous artist Eileen Harrison painted an image in 2005 that depicts Aboriginal women at the Den of Nargun. The caption on the painting reads '[c]oming down towards the river, as you watch, spirits watch over you. The protectors of the boulders represent beauty of the forest. The black trees are blackened through fire' (Culture Victoria, 2010). The same image features in the book *Meerreeng-An Here is my Country* (Keeler & Couzens 2010: 46). Russel Mullet, a Kurnai man, shares his story about the Nargun: 'me-man-dook grandmother ghee-yin tell dyerrak fear tukai child bitja fire ngarwu listen wan-gin-mirri hear' meaning 'after dark, we would sit around the fire and the old people would tell stories about Mrarts and Doolagahs and Narguns. In our imagination, they were something like Big Foot, big hairy men, wild men, and the Narguns were their wives' (Keeler & Couzens, 2010: 46).

It is proposed that the Den of Nargun as a cultural and natural tourism attraction has not reached Butler's period of decline or rejuvenation. Rather, it is still in a period of stagnation. Firstly, as stated by Butler (1980) decline occurs when the nucleus becomes less attractive to the tourist. There is no statistical evidence that supports a decrease in site attractiveness. Upon visiting the Den of Nargun in January 2013, the picnic area appeared to be well maintained with the facilities all above average. It is argued that if the Den of Nargun was experiencing a period of decline, the facilities and structures would not be maintained to the perceived good standard. Conversely, it is also believed that the Den of Nargun is not experiencing rejuvenation. If rejuvenation was to occur, it would be a result of clear and well planned strategic management and marketing plans (Butler, 1980). Given that the most current management plan is from 1998 there is little evidence to suggest that dramatic interventions at the site have occurred in recent years. Therefore, it is argued that the site and its features are in a period of stagnation. The only known form of tourist focussed rejuvenation is the restructuring of the Bataluk Cultural Trail in 2011 (ABC Melbourne, 2011). Overall the site is continuing to be maintained at a reasonable standard contributing to the argument that the Den of Nargun has not yet entered rejuvenation or decline.

5.5 Fifth Phase: Social Reproduction

For MacCannell (1976), the final stage of sight sacralization is social reproduction which occurs when groups, cities, and regions begin to name themselves after famous attractions. In the case of Den of Nargun it has not been possible to locate any instances of social reproduction.

5.6 Tourism at the Den of Nargun

The Den of Nargun has a long history of tourism for both natural and cultural reasons. Since its European discovery by explorer and natural scientist Alfred William Howitt in 1875 it has progressed into a tourist site with the first large influx of tourists in the 1930s (Australian Government, 2011; Catrice, 1996). The Den of Nargun continues to attract over 50,000 visitors annually. Its successful tourism history can be attributed to its aesthetic and Indigenous values and history. The conglomerate cavern formation of the cave is believed to have been formed over 36 million years ago and is the largest cave of this formation on record (Harvey, 2007; Department of Environment and Primary Industries, 2012). This unique and picturesque site attracts visitors. The supernatural being believed to inhabit the cave provides the tourist with a unique and enlightening experience. Table 5.1 outlines critical moments in the history of tourism at the Den of Nargun.

Tab. 5.1: Den of Nargun: a timeline

Year	Event
1875	European discovery by Alfred William Howitt.
1884	Local selector J.D. Stocks 'discovers' the Den of Nargun.
1886	Journalist Julian Thomas aka 'The Vagabond' publishes account of a visit to the Den of Nargun.
c. 1900	Discovered by the Morrison Brothers, local landholders.
1904	Morrison Brothers return to site with Professor Stanley Porteus
1923	Noted bushwalker, R.H. Croll documented the Den of Nargun and its location in newspaper accounts and in his guidebooks -'The Open Road' (1928) and 'Along the Track' (1930).
1926	Charles Daley and a party of the Field Naturalists Club of Victoria visit the Den of Nargun.
1930	Large influx of tourists.
1938	Den of Nargun and the area of Woolshed Creek were declared a sanctuary.
1963	Den of Nargun registered as part of the Glenaladale National Park.
1980	Den of Nargun registered as an historic site by the Australian Heritage Database.
1986	Den of Nargun became part of the Mitchell River National Park.
1998	Parks Victoria management intervention and site development. Estimated tourists to the Mitchell River National Park were 56,000 annually.
2007	Harvey published the Conservation Analysis Report: The Den of Nargun
2010	Mitchell River National Park joint management plan between Parks Victoria and Gunaikurnai Land and Water Corporation.

According to Butler's (1980) tourism area life cycle model, a tourist site will experience either rejuvenation or decline following stagnation. The Den of Nargun is still in a period of stagnation and it may enter a period of decline if there is not any further tourism management interventions initiated by the managing bodies Parks Victoria and Gunaikurnai Land and Water Corporation. There are also concerns about protecting the nucleus from tourism impacts. Harvey's (2007) report highlighted some key areas for improvement in order to sustain tourism at the Den of Nargun. Firstly, the interpretative signage at the Den of Nargun Picnic Area and along the track to the den itself is out-dated. Harvey (2007: 6.5.3) writes 'the signage at the den needs to be updated and expanded, with greater emphasis given to the importance of this special place'. This statement was confirmed by the first named author during a recent visit to the Den of Nargun where there was only one sign with limited interpretation located sixty metres from the nucleus. Whilst the facilities at the Den of Nargun Picnic Area are well maintained, the walking track has become heavily eroded in sections making it difficult to navigate. The current structure of the walking track limits accessibility for people with young children or low physical capabilities. This needs to be considered because if the track continues to erode the ability to view the nucleus will be severely compromised.

There are also concerns that there is a lack of intervention to minimise tourists entering the cave (Harvey, 2007). Park Ranger Yasmin Aly, has noted 'We have a walking track to the Den and we're trying to manage people wanting to sit in the

Den which they shouldn't do because of the cultural significance. I'd like to see more input from the community, perhaps...developing a working group with the local women in the indigenous community so we can refer to them when we have issues such as the Den of Nargun access' (Environment Victoria n.d.). Despite the majority of stalactites formed on the exterior of the den and interior of the cave being broken off by tourists, there has been no intervention such as the construction of a grille or some other barrier that would act as a physical inviolate belt to reduce degradation and tourism impacts. Parks Victoria (1998: 16) stated that they aim to '[m]inimise the impacts of visitor and management activities on significant geological features'. Furthermore, The Department of Primary Industries (2012) expresses concerns about the tourism impact on the aesthetic value of the Den of Nargun. However, the nucleus remains unprotected with no deliberate attempt to install an inviolate belt. Overall, these notions need to be considered if the sight/site is to continue to provide positive tourism experiences in future.

5.7 Conclusion

The Den of Nargun, set in the Mitchell River National Park in Gippsland, approximately 50km from Bairnsdale, has had a successful tourism history since its European discovery in 1875. This study has attempted to document the development of the site utilising the theoretical constructs developed by MacCannell (1976), Butler (1980) and Gunn (1994). MacCannell's progressive development of attractions amalgamated with Butler's tourism area life cycle model has provided systematic insight into the different phases which occurred since the Den of Nargun's discovery to its current maturation. The investigation into the tourism history of the Den of Nargun was further enhanced through application of Gunn's spatial model which provided insight into the contextual setting of the nucleus and its surrounding environment. Potential areas for site improvements presented by Harvey (2007) were supported by photographic evidence and anecdotal analysis at the site in January 2013.

Although 'discovered' by Europeans in 1875, the Den of Nargun did not begin to experience significant numbers of visitors until the 1930s largely through the agency of bushwalkers and field naturalists. Although the site was declared a 'sanctuary' in 1938, formal site protection did not take place until November 1963 when the private company, Australian Papers Manufacturers Ltd, donated 163 acres that included the cave that saw the formation of the Glenaladale National Park. In 1986 the Glenaladale National Park was merged with adjoining land and renamed the Mitchell River National Park. Today, the site is managed by Parks Victoria and the Gunaikurnai Land and Waters Corporation.

References

ABC Melbourne. (2011). *Breakfast with Gerard Cullinan: Bataluk Cultural Trail.* Retrieved 8 December 2012 from http://blogs.abc.net.au/victoria/2011/12/bataluk-cultural-trail.html

Australian Government. (2011). *Place Details: Den of Nargun, Walpa, Victoria.* Retrieved 8 December 2012 from http://www.environment.gov.au/cgibin/ahdb/search.pl?mode=place_detail;place_id=103369

Australian Heritage Places Inventory. (N.D.). *Den of Nargun.* Retrieved 8 December 2012 from http://www.heritage.gov.au/cgi-bin/ahpi/record.pl?RNE103369

Bairnsdale Advertiser August 1962; 26/1/1884

Bataluk Cultural Trail. (N.D.a). *Explore the Bataluk Cultural Trail.* Retrieved 8 December 2012 from http://www.batalukculturaltrail.com.au/index.php

Bataluk Cultural Trail. (N.D.b). *Bataluk Cultural Trail brochure.*

Billis, R.V. & Kenyon, A.S. (1974). *Pastoral Pioneers of Port Phillip.* Melbourne: Stockland Press.

Butler, R.W. (1980). The concept of tourist area cycle of evolution: Implications for management of resources. *Canadian Geographer*, 24:5-12.

Calder, J. (1990). *Victoria's national and state parks.* Melbourne: Victorian National Parks Association.

Catrice, D. (1996). *Victoria's Heritage: Mitchell River National Park.* Melbourne: Parks Victoria.

Clark, I.D. (2007). The abode of malevolent spirits and creatures – caves in Victorian Aboriginal social organization. *Helictite*, 40(1):3-10.

Croll, R.H. (1928). *The Open Road in Victoria: Being the Ways of Many Walkers.* Melbourne: Robertson & Mullens.

Croll, R.H. (1930). *Along the track.* Melbourne: Robertson & Mullens.

Culture Victoria. (2010). *Indigenous Culture: Meerreeng-an Here is my Country Den of Nargun.* Retrieved 7 December 2012 from http://www.cv.vic.gov.au/stories/meerreeng-an-here-is-my country/10699/den-of-nargun/

Daley, C. (1927). The Mitchell Gorge Excursion. *Victorian Naturalists Journal*, xliii: 297-302.

Department of Environment and Primary Industries. (2012). *8322-5 Den of Nargun.* Retrieved 9 December 2012 from http://vro.dpi.vic.gov.au/dpi/vro/egregn.nsf/pages/eg_lf_sites_significance_8322_5

Environment Victoria. (N.D.). *Yasmin Aly: A River Love Affair.* Retrieved 7 December 2012 from http://environmentvictoria.org.au/content/yasmin-aly

Etheridge, R. (1893). Note on an Aboriginal Skull from a Cave at Bungonia. *Records of the Geological Survey of New South Wales*, 3(Pt. 4):128-131.

Field and Naturalist Club of Victoria. (2012). Retrieved 12 December 2012 from http://www.fncv.org.au/ welcome.htm

Gippsland Pictorial News, December 1962

Google Maps. (2013). *Get Directions: A) Glenaladale, Victoria B) Mitchell River National Park, Cobannah, Victoria.* Retrieved 14 January 2013 from https://www.google.com.au/maps.

Gunn. C. (1994). *Tourism planning: Basic, concepts, cases* (3rd ed.). Washington: Taylor and Francis.

Harvey, P. (2007). *Conservation Analysis Report: The Den of Nargun.* Bairnsdale: East Gippsland Institute of TAFE.

Hartnett, S. (2008). *The ghost's child.* Camberwell: Penguin Books.

Howitt, A.W. (1875). Notes on the Devonian rocks of North Gippsland. *Third report of progress, Geological Survey of Victoria*, 181-249.

Howitt, M. (1971). *Come Wind, Come Weather: A Biography of Alfred Howitt.* Carlton: Melbourne University Press.

Keeler, C. & Couzens, V. (Eds) (2010). *Meerreeng-an Here is my Country: the story of Aboriginal Victoria told through Art.* Melbourne: Koorie Heritage Trust.

McLean, A. (1866). *Lindigo.* Melbourne: H. T Dwight.

MacCannell, D. (1976). *The tourist: A new theory of the leisure class.* New York: Schocken Books.

Massola, A.G. (1962).The Nargun's Cave at Lake Tyers. *Victoria Naturalist*, 79(5):128

Massola, A. (1969). *Journey to Aboriginal Victoria*. Adelaide: Rigby Limited.

Parks Victoria. (1998). *Mitchell River National Park Management Plan.* Retrieved 14 December 2012 from https://www. parkweb.vic.gov.au/__.../Mitchell-River-National-Park-ManagementPlan.pdf

Parks Victoria. (2011). *Park Notes: Mitchell River National Park.*

Parks Victoria. (2012). *Park Notes: Mitchell River National Park.*

Parks Victoria. (2013). *Mitchell River National Park.* Retrieved 9 January 2013 from http://parkweb.vic.gov.au/explore/parks/mitchell-river-national-park

Pascoe, B. (1986). The slaughters of the Bulumwaal butcher. *Anthology of Australian Literature.* Australia: Allan & Unwin

Pearson Australia (N.D.). *Unit 16: Den of Nargun.* Retrieved 9 January 2013 from http://www.pearson.com.au/media/297909/pdf-literature-word-up-grammar1-u16-page-proofs.pdf.

Pepper, P. (1980). *You are what you make yourself to be: the story of a Victorian Aboriginal family 1842-1980.* Melbourne: Hyland House.

Porteus, S.D. (1951). *Providence ponds: a novel of early Australia.* Bairnsdale: James Yeates and Sons.

Red Bubble. (2012). *Den of Nargun by Dave Callaway.* Retrieved 12 December 2012 from http://www.redbubble.com/people/davecall/works/9170715-den-of-nargun?p=greeting-card

Seddon, G. (1989). *Searching for the Snowy.* Sydney: Allen & Unwin.

Smyth, R.B. (1878). *The Aborigines of Victoria.* 2 Vols. Melbourne: Victorian Government Printer.

The Argus 3/4/1886; 2/6/1923

The Weekly Times 16/5/1917

Wakefield, N.A. (1967). *Naturalist's Diary.* Melbourne: Longmans.

Wheeler, G. (1991). *The Scroggin Eaters: a history of bushwalking in Victoria.* Melbourne: Federation of Victorian Bushwalking Clubs Inc.

Wrightson, P. (1973). *The Nargun and the stars.* Richmond: Hutchinson. [1988, Hawthorn: Hutchinson].

Yeates, R.S. (1965, March 22). Mr. E.D, Gill, assistant director, National Museum of Victoria, Melbourne. Bairnsdale Historical Society.

Images

Culture Victoria. (2010). *Indigenous Culture: Meerreeng-an Here is my Country Den of Nargun.* Retrieved 12 January 2013 from http://www.cv.vic.gov.au/stories/meerreeng-an-here-is-mycountry/10699/den-of-nargun/

Fig. 5.1 :The Den of Nargun (Sharnee Sergi, 2013)

Fig. 5.2: '"Ngrung-A-Narguna" A cave in Dead Cock Creek Mitchell River' A.W. Howitt (1875: facing page 220).

Fig. 5.3: A shooting Party at the Den of Nargun in 1905 (Source: Courtesy of Department of Environment and Primary Industries, Victoria)

Fig. 5.4: Den of Nargun picnic area (Sharnee Sergi, 2013)

Fig. 5.5: *The Nargun and the Stars* (1974) cover

6 Hanging Rock Recreation Reserve by Stephanie Skidmore and Ian D. Clark

This chapter presents an historical analysis of the evolution of tourism at Hanging Rock Recreation Reserve. It shows how Hanging Rock evolved from being a 'special' place of local Aboriginal clans at the time of European settlement into a significant natural/cultural tourism attraction. Hanging Rock Recreation Reserve is some 80 kilometres N.N.W. of Melbourne and lies immediately north of Mount Macedon. It is managed by the Macedon Ranges Shire Council. In terms of its physical extent, the acquisition of 22 hectares in 1993 increased the size of the reserve to 88 hectares, although the rock itself only covers approximately 9 hectares, the remaining area is comprised of the racecourse, picnic areas, and car park. Five Mile Creek, a tributary of the Campaspe River, flows through the southern portion of the reserve. With its distinctive geological formation the rock rises some 100 metres above the surrounding plain.

> The rock formation is a fine example of a volcanic mamelon created millions of years ago in which lava in the central plug spilled out to form a conical shaped hill. Other nearby examples are Camel's Hump to the south on Mount Macedon and Brock's Monument to the east. All three are made of solvsbergite, a form of trachyte found in Solvsberget, Norway and Abyssinia [Ethiopia]. As Hanging Rock's magma cooled and contracted it split into rough columns that weathered over time into pinnacles that can be seen today (Gisborne and Mount Macedon Districts Historical Society, 2012: 3).

Hanging Rock is situated within the traditional country of the Woiwurrung or Wurundjeri Aboriginal people and fell within the estate of the Gunung willam baluk (Creek dwelling people) whose country centred on Mount Macedon (Clark, 1990). Indigenous information about the rock's use in pre-European times is sparse, but it is believed to have been used for intertribal meetings and male initiation ceremonies (Gisborne and Mount Macedon Districts Historical Society, 2012: 6). Archaeological surveys in the district have established a human presence in the area 36,000 years ago (Loder & Bayly, 1993: 10). One of the last initiation ceremonies was held there in November 1851 by a Wurundjeri elder from the Templestowe area (Poulter, 2011: 32-3). Today the Wurundjeri people maintain strong cultural and spiritual ties with the area (Gisborne and Mount Macedon Districts Historical Society, 2012: 6).

6.1 First Phase: Sight Sacralization and Naming 1844–1860

In terms of MacCannell's (1976) first phase in the development of attractions that of 'sight sacralization' or 'naming', one Aboriginal name 'Anneyelong', and at least six European names 'Mount Diogenes', 'Diogenes' Head', 'Diogenes Monument',

'Dryden's Rock', 'Dryden's Monument', and 'Hanging Rock' have been recorded for this site.

When William von Blandowski sketched the formation in 1855-1856, he called it 'Diogenes Monument' and listed directly underneath this label 'Anneyelong,' which is presumably its Indigenous name (see Figs.6.1 & 6.2). The meaning of 'Anneyelong' is not known.

Blandowski's (1855) geological observations were published in *The Star* (22/9/1855). They are important as they are the earliest account of the rock:

> "Dryden's Monument" appears to us one of the most remarkable spots in the country- at least of those with which we met while traversing an area of more than seven hundred miles westward and four hundred miles northward. A careful and minute description of it, accompanied with good drawings and models would justly deserve the attention of all geologists. An almost perpendicular wall of dolorite, of a deep sombre hue, rising above the most lofty of the trees, imparts a strikingly picturesque character to the view. Approaching the Monument, the interest increases at every step, and we meet with a beautiful variety of rapidly changing scenery. At the foot of the monument, about [a] thousand pyramidal columns rise in bold relief from the surface, giving to the hill, which is about a mile in circumference at the base, a kind of gigantic porcupine appearance. These columns vary from about fifteen to thirty feet in diameter at the base, and from thirty-two to one hundred feet in height. The formation of this hill is the result of subterraneous agency, and has evidently taken place at two different periods. A naked semicircular hill was formed during the first period; and while the surface was scarcely hardened, and yet in a plastic state a second eruption occurred, and resulted in the production of the peculiar pyramidal columns already mentioned.29

Mackay (2011: 121) notes that Blandowski wants to be geologically informative, but is 'carried away by the need to impart to the reader an intensity of his sense of being in the presence of something strange, singular and sublime in the natural world'. His choice of words intimates the latent expectation of a sublime experience, the word 'grand' and 'grandeur' often being synonymous with 'sublime' in eighteenth- and nineteenth-century writings.

Elsewhere, Blandowski (1855: 51) noted that 'at the head of Five Mile Creek, is a remarkable hill called Diogenes' Mount, commonly known to the colonists as "Dryden's Monument", a name singularly inappropriate, being the cognomen of a settler in the neighbouring district'.

Mackay's (2011: 121) analysis of Figure 6.1 reveals that the engraver deliberately accentuated danger and menace in the engraving:

> Diogenes' Monument, also known as Dryden's Mount, and today called Hanging Rock, projected a mysterious and threatening image in nineteenth century representations of the site. The so-

29 Blandowski (1855: 51) described Dryden's Monument as a 'singular conformation' and one of the most remarkable spots in Victoria, if not in the whole of Australia. The approach to the monument 'presents a scene of most imposing grandeur'.

> called 'rock' is a small mountain near Mt. Macedon in the Woodend district of Victoria, distinguished by its dense mantle of stupendous rock crags and outcrops. In Blandowski's engraving, titled Diogenes Monument, moonlight illuminates a surreal scene in which unbelievable rock forms locate the sense of unease generally felt by visitors to the location. The dark shadows and menacing shapes are barely alleviated by the campfire of an Aboriginal group on the hillside. Notes accompanying a letter to Blandowski from Germany suggest that the engraver intended to convey the potential danger in nature-that he saw Australia as a country close to a time when primordial forces ruled the earth.

Heckenberg (2011: 383) notes that the distant view expressed in this engraving fulfils Blandowski's comment that minute description would interest geologists, especially with its detail of the rocks and vegetation, but the added drama provides interest for all viewers. 'Two Europeans exclaim in wonder in the foreground as they regard the hill, columns illuminated by the light of the moon. In this image Aborigines are at home, camping in the middle ground while two other mysterious figures observe them on the right'. For Heckenberg (2011: 385) Blandowski's scientific vision is concerned with 'the specifities of the land, such as details of its geology, vegetation and weather and also its indigenous peoples and their activities'. In this his images contribute successfully to 'Humboldt's desired animated and picturesque representation'.

Mackay's (2011: 122) analysis of Figure 6.2 is that it is invaded by a sense of threat 'where forked lightning strikes the ground and throws into relief the extraordinary outlines of three of the mountain's cone-shaped rocks. An Aboriginal figure reels back in horror at the spectacle; his arm raised in astonishment: the classic convention to signify he is in the presence of the sublime'.

> THE accompanying engraving represents a scene presenting features of a most remarkable and picturesque character. Dryden's Rock, Mount Macedon, is in the county of Dalhousie, Victoria, about three miles from Woodend, As will be seen, the view is peculiarly Australian in its character, and the principal object represented may be taken as a striking specimen of those numerous isolated masses of rock which at once arrest the traveller's attention, and are to be found in several districts of both New South Wales and Victoria. This singular natural feature consists of a lofty detached mass of light coloured porphyritic felspar trap, divided by numerous joints into rude columns from 30 feet to 40 feet in height and exhibiting a concretionary structure, well displayed in large circular and funnel-shaped cavities, which are abundant throughout the mass. At the base of Dryden's Rock, on its south-east side, there is a fine spring of water, which forms the head of a creek known as the Five-mile Creek.

Figure 6.3 shows Indigenous people sitting around a fire around the base of Dryden's Rock.

Local historian Marion Hutton (1991) has discussed the site's nomenclature. The first European name that was conferred on the rock was Mount Diogenes, which is often incorrectly attributed to Major Thomas L. Mitchell's 1836 Australia Felix expedition. However, as research by Hutton has revealed, there is no evidence to support the Mitchell origin. The earliest use of the name Mount Diogenes is on surveyor Robert Hoddle's map of 1844 (Hutton, 1991: 3). Hutton considers it likely that Hoddle, wishing to keep the continuity of Mitchell's Greek names (such as Mt. Macedon, Mt. Alexander, and Campaspe River), was responsible for the name. Diogenes was a fifth century philosopher who reputedly lived in a tub at Corinth; when Philip of Macedon threatened to capture the city, Diogenes rolled his tub back and forth in defiance (Blake, 1977: 184).

The early settlers, however, called the rock 'Dryden's Rock', in reference to the fact that from 1837 the rock formed part of Edward Dryden's 'Newham' pastoral station, which he named after his natal place in Northumberland, England. In 1855 Blandowski used the name 'Dryden's Monument'. When Richard Daintree photographed the rock in 1858-1863, he used both names Dryden Rock and Hanging Rock. Mount Diogenes was still in use in 1874 as the name appears in a wood engraving of 'Holiday rambles at Mount Macedon' by Samuel Calvert published in the *Illustrated Australian News for Home Readers* (25/2/1874). The name Mt Diogenes with Hanging Rock in brackets appears on a 1912 Geological Survey map (see Gisborne and Mount Macedon Districts Historical Society, 2012: 4) which confirms that the former is the official toponym for this site. A correspondent to *The Argus* (17/10/1917) noted that 'Hanging Rock, or Mount Diogenes (its geographical name), does not 'hang', though I understand there is a comparatively small mass of rock which is so curiously balanced that it puzzles the onlooker as to how it maintains its position'.

Hutton (1991) considers Hanging Rock is a nick-name that dates from the 1850s, and is believed to derive from the enormous boulder lying suspended and wedged between two others (this is confirmed by H. Follett in the *Williamstown Chronicle* 25/1/1896) (see Fig.6.4). The source of this nick name is unknown but it probably originated from early settlers and/or their employees. Hanging Rock is therefore a descriptive microtoponym for a geographical feature that is seen in Figure 6.4, and not the official name for the mountain which remains Mt Diogenes. The *Australian Town and Country Journal* (25/5/1889) took issue with the name, asserting that 'the name "Hanging Rock", however, as commonly applied to the whole mount, is an evident misnomer. It should be 'the Hanging Rocks,' for at nearly every 100 yards or so, tremendous boulders, many tons in weight, hang, apparently ready to topple upon the onlookers at a moment's notice'. A columnist in *The Canberra Times* (1/7/1989) reported that 'The proper name of Hanging Rock near Woodend, Victoria, is Mt Diogenes'.

Edward Dryden and Charles Peters first occupied the site in 1837 when they took up the 'Newham run', some 25,000 acres adjoining Woodend. In 1848 Dryden was sole lessee of the run. In 1857, a Kyneton solicitor, Thomas Lloyd James, purchased 170 acres which included the rock. Three years later James sold 100 acres including the rock to Alexander Archer, a Kyneton bank manager, and the remaining 70 acres to William Adams (Hutton, 1991). It was during Archer's ownership that we have the first newspaper account of an organized picnic held at the rock organised by Kyneton tradesmen on Boxing Day 1864.

When more and more people started to visit the rock, for enjoyment and tourism purposes, various rocks formations were soon named. The micro-toponymy at Hanging Rock includes the Post Office, where passers-by tossed small stones into a hollow place; The Lovers Leap: a rock jutting out from the cliff; Queen Victoria's Monument: a replica of Queen Victoria in her robes; McDonalds Look-out where McDonald, a Bush Ranger, was able to look out over the North-East between two rocks, while Morgan, another Bush Ranger, had his look-out Westerly from a hollow rock with an opening in it. The boys of that time had one Rock named The Cuss-Cuss Rock where the little native cats would spit at passers-by. The Squeeze was a tunnel from the top above Hanging Rock to the base. It meant clambering over and crawling through small spaces, until safely reaching the ground level (Macedon Ranges Tourism, n.d). Other names include 'Alligator's Mouth', 'Ticket Office', 'Duke of Wellington', and 'The Sphinx'. The Friends of Hanging Rock have produced a map and a guide with the current names.

6.2 Second Phase: Framing and Elevation 1864–1884

The second phase identified by MacCannell (1976) in the evolution of attractions is 'framing and elevation' which he argued results from an increase in visitation, when demand requires some form of management intervention, whereby the sight is displayed more prominently and framed off. In the case of Hanging Rock this phase began in 1864 with the first formal picnic to the rock and ended in 1884 with the Victorian Government purchasing Hanging Rock and reserving it as Crown Land.

The earliest recorded account of the rock being used for leisure purposes was on Boxing Day 1864 when a picnic was organized by the tradesmen of Kyneton. Participants were asked to bring quoits, bats and balls and were told ale, lemonade and ginger beer would be provided at the ground (Hutton, 1991: 4-5). On the day 'a long string of conveyances of every description from the buggy to the spring cart filed off along the road preceded by the band and accompanied by an equally large proportion of equestrians' (Hutton, 1991: 5).

In 1869, Alexander Archer sold the 100 acres (including the rock) which he had purchased in 1861 to William Adams for £550 (*The Argus*, 5/4/1869). Adams now owned 170 acres. In 1866 he had acquired a half-acre block on the north-west corner of the Rock, where the Coach Road joins the Woodend-Lancefield Road and built the Hanging Rock Hotel. The following year he bought some 22 acres south of the creek.

Adams set about developing the area as a pleasure resort. He placed a weir across the creek to form a lake suitable for small rowing boats and other small water craft as we learn from the *Kyneton Guardian* of 1869 – 'the neighbouring publicans got up a full programme of sports, among them the driving of a pair of geese harnessed to a tub, across the small lake' (Hutton, 1991: 5). Adams also constructed a road that went almost to the top of the rock. It 'started at the foot of the rock on the western side and climbed around the southern side into the large open area at the top of the Rock just below the final peak. It was wide enough for a cab to be driven up with safety. Ornamental trees were planted and beds of flowers laid out' (Barned, 1985: 32). There were also 'foot races, jumping in sacks and other amusements not forgetting music to tempt those anxious for a dance'. The entrepreneurial Adams organised sports meetings and together with the publican of the Garden Hut Hotel, Robert Walsh King, he ran the booths for refreshments and for music and dancing. Adams and King were granted temporary licences for the holiday periods and the publicity in the *Kyneton Guardian* was quite imaginative, as revealed by the following jingle that appeared in successive issues in 1869 (Hutton, 1991: 5):

SPORTS – SPORTS – SPORTS

Where are you going to spend Christmas Day? At the HANGING ROCK RESERVE, as at King's Booth there is sure to be plenty of Music and Dancing.

Where are you going to spend Boxing Day? At the HANGING ROCK RESERVE, as the Doctor ordered Father to take us all to hear the UPSON'S CHRISTY MINTRELS at King's Booth and says it will be better than all his medicine.

Where are you going on New Year's Day? TO the HANGING ROCK RESERVE, as Everybody and two or three other people are going to hear the Christy Minstrels at King's Booth.

In 1870 some 96 acres of land immediately west of Hanging Rock, which had never been privately owned, were gazetted by the Victorian Government as a reserve for recreation and water. The Woodend, Newham and Rochford Road Board (precursor of the Newham Shire Council) was instrumental in this reservation when it learned that moves were afoot to sub-divide and sell the land.

> Adams must have got into financial difficulties because we find him taking out a mortgage and in 1871 he sold all but the portion with the Rock to William Anderson who already owned land south of the creek. By 1873 Adams was prepared to sell the Rock to the Government, the Newham Shire Council having initiated moves in that direction. He was willing to sell at the price he paid (£5-10s per acre) plus the value he placed on his improvements (£260). The surveyor sent to inspect reported it to be [a] 'most preposterous valuation' but he added 'I must do the owner the justice of stating that the improvements that have been made in excellent taste and render the access to the summit of the Rock much easier and the place altogether more charming' (Hutton, 1991: 7).

However, the Government considered Adams' asking price too high, and the sale did not proceed. Adams continued to own the Rock and cater for picnics and sports. By this time people were coming from further afield, places such as Melbourne, Sandhurst (Bendigo), Kyneton, Castlemaine, Echuca and other parts of the colony. Visitors came on all public holidays – Christmas, Boxing Day and New Year's Day – when the railways ran special holiday trains. Horse races were introduced to the impromptu programme, the local farmers competing for whips and saddles. Hutton (1991: 7) considers 'It is an exaggeration to say that Adams laid out a race course for the purpose; he cleared a rough track among the timber for typical bush picnic races and he provided a pavilion for dancing on the flat near where the sports were held'.

On 28 December 1871, *The Argus* published an account of a trip by a correspondent and their friend to Hanging Rock on Christmas Day. Taking the train from Melbourne to Macedon, they ascended Mount Macedon and then walked to 'another part of the same group, called by the map-makers Mount Diogenes, by residents in the district 'the Camel's Hump', and by old hands in the earlier days the Peak.[30] The last name is the best and most expressive, as, although not higher than the ground we stood upon, it is the real central peak of the Mount Macedon block'.

> The Hanging Rock lies about four or five miles from Woodend, about due north from Mount Diogenes, whence we had over looked it. It is a hill of about 300ft. high, and of irregular conical shape. A little way up from its base it is surrounded by a threatening looking rampart of grey trap rook, which in some places rises to a perpendicular height of 40ft. or 50ft. The massive wall of rock is fissured by crevasses to its foundation, and a rude columnar structure is thus imparted to it, the separate blocks at their tops weathering off into ragged pinnacles by which the ruined fortress character of the whole is much enhanced. These blocks are again divided in places by horizontal bed lines, and in all of these joints there is sufficient moisture to nourish the tree ferns, musk bushes, and various rock plants, which give a beautifully verdurous appearance to the old grey ruin. A path winds up the hill, up the grassy slope at, the foot, up between two towering rock walls, beneath an enormous boulder which lies jammed in between them, forming the cyclopean gateway of the fortress, up the tree-covered slope, which is found again within the outer rampart, till it is finally lost amidst the various rock masses which crown the top of the hill. The Hanging Rock is the great place for picnic parties for the whole district. On holidays people come to it from Kyneton, Kilmore, Lancefield, Woodend, and all the neighbouring places, and custom to them never stales its infinite variety. When we reached it, about 2 o'clock in the day, we saw scores of vehicles standing about amidst the trees, hundreds of saddle horses hitched up to the fences, many picnic groups disposed about here and there on the grass, two dancing booths resounding with music, refreshment tents, crowds of promenaders up and down the rock, men and boys on horseback riding harum-scarum races through the timber, here a group watching pole-vaulting, or getting up a footrace, or playing quoits–altogether a great spontaneous gathering of many hundreds of people in the most delightfully picturesque spot to be found, and all evidently enjoying themselves to the last degree. The great holiday musters at the Hanging Rock have a character of their own. There is no organisation, no general arrangement; each party provides for itself, and acts as it pleases. And I may say that I never saw in Victoria, or anywhere else, equally large gatherings where there is so little disorder, irregularity, or unpleasantness of any kind as at those.

30 The correspondent is confused here – Mount Diogenes and Camel's Hump are distinct and separate formations.

After dinner we nearly all went up the Rock, stooped under the massive gateway, climbed up the last slope, and sat crowded, about 20 of us, on one of the highest rocks, over-looking the country around us with our glasses. I at once meet the deprecatory feeling of the reader by repeating my declaration that I don't for a moment meditate writing any more description. I only mention the fact that we spent a good deal of time watching the view all glowing with light and life in the bright sunshine. To many of us the prospect was a familiar one. We knew the occupants and the histories of many of these homesteads scattered there in the agricultural country outspread beneath us. To those to whom it was new and strange there was enough in the view itself to make it a delight. Having stayed there for a time, we were led away by a proposal made by some of the frequent visitors of the rock, that we should go down by one of the crevices. Of course, seeing there was no escape, we at once agreed, and said it was the thing above all others that we wished to do. However, we did not contest priority, and gracefully allowed some of the others to go first. When it came to my turn to descend, I followed the example of the others, and, turning my back to the way we had to go, lowered myself into a very narrow orifice between two rocks, letting myself down as easily as possible till I reached the hard bottom. I then stood in a narrow fissure that had some time opened in the enormous mass of rock, and remained as a joint between them of a few inches in width, and of what extent upwards and downwards you could not see. The path beneath your feet was steep and slippery; the sides at points closed in upon you, as though you must be inevitably jammed a step or two further on; the walls seemed to close in above you, and to exclude the light and make the air most oppressive. There was in your favour the knowledge that others were going on ahead of you, although how they did so you could not see, and a pleasing stimulus was supplied to your squeeze downwards by the certainty that, in the worst case, it was hopeless to attempt getting back, And so in one way or the other, by holding your breath, narrowing your chest, and edging yourself through the tightest pinches, and by gradually working downward, you at length emerged into the daylight, and found yourself in a bed of tree ferns and musk trees at the foot of the great rock rampart I have before described.

[T]o all people who wish on a holiday or on a day they have set aside for an excursion to escape from all the conventional scenes of holiday resort; who are wearied of Brighton beach and Mordialloc, and even don't wish to climb up Ferntree Gully anymore; who desire to find a place of interest, variety, picturesque beauty, fresh and unworn, with all the charm of the genuine country still about it. To all such people I wish to give my strongest recommendations to pay a visit to the Hanging Rock. They may have prejudices against Australian scenery; they may believe that it possesses no elements of beauty, diversity, grandeur; that it displays nothing but the unbroken monotony of endless forest or endless plain; that it is parched in early summer, wanting in freshness, and entirely destitute of the picturesque. Well, let them go to the Hanging Rock, and reconsider their judgment after they return. And whatever their judgment as to this may be, I am still sure of this, that they will agree that at the least as a place of picnic resort there is none to rival this one.

In saying this I am perhaps showing an ill return for the welcome I received. Should the Melbourne people ever flock to the rock in one-tenth of the numbers that they do to places not more accessible and not to compare with it in attractiveness, it would be soon spoiled as a gathering-place for the local residents who now visit it so frequently. But having so much enjoyed my own trip, I wished to point out to the holiday-keepers of Melbourne what points of interest they leave altogether unvisited; and while they complain of want of variety, what varied resources they have which they wholly neglect (*The Argus*, 28/12/1871).

The Illustrated Australian News for Home Readers (23/4/1872) published an image (see Fig.6.5) of the approach to Hanging Rock. It is of interest as it shows tourists on various points of the rock.

In 1874, the *Illustrated Australian News for Home Readers* published another article on the Hanging Rock, and was critical of the fact that the rock was in private hands:

> It will be hardly credited, but this piece of rich natural magnificence, that belongs of right to the whole people of the colony as a sight to look upon and marvel at, has been bartered away by a blundering Lands department for a few paltry pounds; and the Government of the day are too parsimonious to re-purchase it, though it can be re-bought for a very reasonable amount. It is hardly necessary to say that the place is resorted to by visitors in search of natural beauties and by picnic parties' innumerable at holiday times. About Christmas and New Year the grass is trodden down to the bare earth in all directions. People clamber up the hill in thousands, and explore the labyrinths formed by the complication of rocks that cover the sides of the hill. It is known far and wide throughout the colony as one of the most remarkable natural curiosities we possess in the way of scenery, and yet it is allowed to remain in private hands, though its purchase at the present time is a very easy matter.

In 1876, Adams finally severed all connections with Hanging Rock by selling out to William Anderson, a Melbourne warehouse owner; but Anderson was not interested in promoting a tourist attraction and leased the grounds to successive owners of the Hanging Rock Hotel for this purpose (Hutton, 1991: 7).

> By this time, the sporting farmers of the district, many of them involved in breeding horses, wanted more horse racing and a better place to do it. They thought a nice little race course could be made on the reserve east of the Rock so they worked hard through the winter of 1877 to achieve this result, then published the sports programme for New Year's Day 1878 in the newspapers. ... On New Year's Day 1880, the Hanging Rock Cup made its appearance on the programme- 'Of 15 sovereigns, twice round the course and a distance'. It seems fitting that the winner of the first Cup was Edward Dryden's aged horse 'Commodore,' though it has to be admitted there were only two horses in the race (Hutton, 1991: 7,9).

On 17 February 1877, *The Australasian Sketcher with Pen and Pencil* published the following account of 'The Hanging Rock':

> Of all the places of holiday resort for pic-nic purposes open to the choice of the pleasure-seeker, there is none at once more interesting, picturesque, popular, and readily accessible than the Hanging Rock, which lies some four or five miles from Woodend. Its situation and surroundings are interesting and picturesque. All around its stretch farm clearings cutting the forest into alternate squares, like the white and black squares of a chequerboard. Away to the north and east lie bold hills scattered in groups, or linking themselves into regular chains of ranges, some covered with farm holdings, others dark with the almost unbroken forest. To the south rise high and dark the sombre masses of the Mount Macedon range, and in the middle of the crowning ridge stands up bare and rugged, the rocky, roughly fissured cone of Mount Diogenes.

> The Hanging Rock itself is a hill of rudely conical form, from 300ft. to 400 ft. in height, and crowded in its upper part with an enormous mass of rugged trap rocks, broken and riven and weathered into the most fantastic forms. The shape most affected is the pinnacle, but in places the rock solidifies into a wall of perpendicular precipice, rising to 50ft. or 60ft. In other parts it resembles the rough cyclopean masonry of an earlier world. The path generally followed up the hill passes under the foot of a frowning, lofty, overhanging cliff, on the broken face of which rock-ferns and musk bushes struggle for life. It then ascends between the walls of an enormous portal, formed by an immense boulder of some scores of tons in weight, lying jammed between the side walls of rock. Beyond this the path climbs up a grassy ascent, out of which stand up strangely-shaped crags, some scooped and hollowed out by the weather into mere shells, others crowned with great toppling boulders, only held in place by vast wedges of rock. And amidst all grow graceful white gum trees, with their clean white stems and green drooping foliage, casting delicate, gently waving shadows on the grey surfaces of the lichen-grown rocks. Above this rises the second cone or citadel of what seems a grey defaced ruin, of such antiquity that all trace of form is lost. The strange shapes of the rocks and the curious alliances of stone and tree everywhere arrest attention. Here there is a singular hollow in a vast crag, out of which something like a wild petrified head of the earlier ages of the world seems to look. Then another rock is split in two, the split suspending and being kept open by great stones, which have fallen into it, and showing large orifices, through which the eye ranges out and gets charming views of the surrounding country enclosed in the rough rock frame. The most difficult and perpendicular part of the outer rampart is riven with deep crevices, which extend down low into the heart of the hill. To go down one of these is one of the correct things to be done by all visitors to the rock, and the journey is worth the squeeze and the difficulty it involves for the sake of the queer sensation given to the passer by finding himself in a mere crack between immense flat walls, formed by an enormous rock, which had once divided, and might— horrible thought – again close.
>
> The Hanging Rock is on holidays frequented by large crowds of people, some of whom come from the surrounding neighbourhood, and some from remote parts of the colony. On last Christmas Day, for instance, there was a large pic-nic party of nearly 100 persons from Melbourne, and two large parties from Sandhurst. But these were nothing in number compared with the numberless parties congregated from the well-peopled districts of Newham, Rochford, and Lancefield, who had driven in buggies and light carts, or ridden or walked to the spot. These were scattered on all the slopes and on every shady spot. Vehicles dotted the hill-sides, and horses were tethered to every convenient tree. A fine clear cold spring rises at the foot of the rock, and affords an abundant supply of water. It is humiliating to add that a place so well suited to serve as a holiday resort to thousands of people, and thus to be a great public benefit to the community, was sold by the stupid short sighted policy of the Government some years ago. The department found the allotment on its plans, and in strict accordance with the genuine of routine offered it for sale. It was sold, we believe, at or near the upset price of £1 an acre, and the effect is, that a place at which very large numbers of people crowd in the course of a year, and which is as a natural object of great picturesque interest, unique in the colony, can only be visited by the permission which, it is only right to say is freely accorded of its proprietor. Of late years he has made a small charge for admission to his paddock, which is, of course, perfectly legitimate. But few can visit the spot without a feeling of indignation at the imbecility of the Government, which, for something less than £100, permitted the alienation of what ought to have been reserved as public property for public recreation and amusement.

At this time there was mounting agitation for the Government to repurchase Hanging Rock for the public. Some of this agitation can be seen in the article in *The Austral-*

asian Sketcher. Anderson allowed public access but made a small charge on public holidays. The site was becoming more and more popular and there were murmurings that it was shameful that it had ever been sold. Fearing that public access might be denied altogether in the future, the people of Newham urged its Council to request the Government to act. With the support of neighbouring shires, a deputation was planned but in the end the Minister, the Hon. Albert Tucker, came to see for himself and he indicated that the Government would be prepared to purchase about 70 acres. After negotiations with Anderson 71 acres were purchased for £1417-12-6, and added to the existing Crown Land reserve and the whole was reserved by Order-in-Council of 25 November 1884 and gazetted a few days later (Hutton, 1991: 9,11).

6.3 Third Phase: Enshrinement 1869 and Beyond

MacCannell (1976: 45) has identified 'enshrinement' as the third phase in the evolution of attractions. 'When framing material that is used has itself entered the first stage of sacralization (marking), a third stage has been entered. This stage can be called enshrinement'. In the case of Hanging Rock this would most likely refer to the facilities and other activities that are available to the visitor at Hanging Rock, such as picnicking and horse racing and the concerts that are staged at the rock, activities that don't necessarily involve any interaction with the nucleus, that is the rock, itself, other than having it provide the location or setting for the activity.

After the Government's reservation of Hanging Rock, in August 1885, a Racing Club was organized. Management of the reserve, however, continued to be a problem for the Newham Shire Council. The Lands Department had ordered the Council to ensure that the local community had unrestricted access to the water in the reserve; however, this led to defilement of the spring and other degradation. In an attempt to manage this, the gates were re-hung and left unlocked. In 1891, regulations were drawn up allowing an admission charge for vehicles on 12 days a year – a vehicle with 1 horse 1/-, 2 horses 2/- and a rider on horse 6d.

By 1891 the racecourse had been allowed to deteriorate with fallen timber making the ground dangerous, so the Council decided to hold a working bee and placed an advertisement in the Kyneton newspapers asking for volunteers (Hutton, 1991:11). The advertisement read as follows:

HANGING ROCK RACECOURSE
MONSTER WORKING BEE

A LARGE number of the inheritance of the district having volunteered to give one or two days' labour on order to clear the racecourse of all dangerous timber now encumbering the ground, it has been decided that a MONSTER BEE SHALL BE HELD ON WEDNESDAY, SEPTEMBER 16th. All those having the popularity, welfare, and improvement of this beautiful and attractive spot at heart are cordially invited to co-operate in this splendid work, and are respectfully requested to assemble on the ground (in the straight) at eight o'clock a.m. Every man to supply his own tools. Several bullock and horse teams have been promised, but the services of more can be utilised.

Remember the day, WEDNESDAY, SEPTEMBER 16th, 1891
By order, JAMES REES, Shire Secretary.
Woodend, August 31, 1891.

(Source: Hutton, 1991: 11)

This advertisement and the community response confirms that the community has always been an important factor in Hanging Rock's history and played a vital role in it becoming a major regional attraction. With the races and other significant events visitor numbers began to increase.

In 1897, *The Australasian* reported on a visit to Hanging Rock:

> A pretty excursion and one within easy reach of Melbourne is to the Hanging Rock, Woodend. There is a service of four trains daily. The time taken to Woodend is two and a half hours. Five miles from Woodend and second only to Macedon itself is the Hanging Rock. This isolated peak rises to a height of four hundred feet above the surrounding country and easy access is gained to its summit by well graded paths along its sloping sides. Its comparatively flat crown is strewn with gigantic boulders, evidently the result of volcanic action in some distant age. Fanciful resemblances are traced in these rocks to well-known personages or animals for us. Thus we have among others, Gladstone's Head; Berry's Head; the Whale's Mouth and the Alligator's Jaws (Gisborne and Mount Macedon Districts Historical Society, 2012: 5).

The Council had formed a committee from within its own ranks, to deal with matters to do with the Rock Reserve such as maintenance, rabbit control and necessary improvements when funds were available. For example in 1896, the dam was repaired to resuscitate William Adams' ornamental lake and in 1900 a carriageway was made up to the first plateau, thereby restoring the latter's old driveway. To mark the end of the 19th century and the inauguration of the Australian Commonwealth, on New Year's Day 1901, triumphal arches were erected over each of the five entrance gates and a 40 foot flagpole was placed in the saddling paddock and another in pride of place on top of the Rock – the socket of which can still be seen on the summit (Hutton, 1991: 15).

The New Year's Day races continued to grow in popularity:

> On New Year's Day 1902, the great day was becoming a popular outing for the family with newspapers estimating attendance of ten to twelve thousand people. Everything on wheels was pressed into service to convey them from the Woodend station. Kyneton was cleaned out of vehicles, so an enterprising Brunswick carrier Herbert Bradford, took to bringing his drays a few days early, to camp on the reserve and spell his horses ready for the hectic day. A special train carried the race horses to Woodend, and landowners allowed them to be led to the reserve across their properties along the route of the Five Mile Creek (Hutton, 1991: 15).

The races on New Year's Day 1911 were particularly special as the Governor and his wife were in attendance along with the Minister for Lands. The following account from *The Age* (2/1/1912) gives a picture of the picnic scene at the Rock the previous day:

> From early morning there had been vehicles on the road. First came vans, spring carts, then wagons lumbering along drawn by horses from the plough. Whole families had to be crowded into the wagon and so had the provisions. Scarcely had the horses been unhitched and tethered and the wagons unloaded, than the drags dashed into the area, whips cracking, drivers shouting, the picnickers laughing. The vehicles crowded the centre of the race course and then overflowed on the other areas of the reserve. Last of all came the hooting motors carrying visitors from Macedon and places more distant. The sun bathed the scene in brilliance as it flicked the caps and colours of the jockeys. The warm light rested on the white table cloths spread unevenly on the ground, covered with poultry, sandwiches, fruits, jams, sweets or other picnic fare. The light made the blue smoke from the big boiler where all came for their supply of hot water, look more blue as it curled amongst the trees. Of the many thousands of people there, only a meagre proportion took the racing as the most important part of the day's outing. The last race was not run until after 6 o'clock and it was not till the late moon rose its shining face above the Camel's Hump that the last picnickers drove slowly away from the grounds.

There were 20,000 people there that day and every year that scene was repeated (Hutton, 1991: 19). The increasing use of the motor car gradually changed the mode of transport and eventually the appearance of the picnic ground on race days. For example, New Year's Day 1925 'was a record year with hundreds of cars parked under trees; dozens of side shows and the band gave a lively atmosphere' (Hutton, 1991: 19).

Other major events at Hanging Rock included sports meetings or Highland gatherings which saw caber tossing, hammer throwing, and foot races. Later a dance floor was erected. Special trains ran from Melbourne to Woodend on Rock Race days and 'Big Four in Hand' drags or coaches which held 40-60 passengers would be driven out to the Rock and later back to the train. At one stage, it was thought a Railway line to connect Lancefield with Woodend via the Hanging Rock would be an advantage and although there were discussions between members of Parliament the train line never eventuated (Macedon Ranges Tourism, n.d).

In terms of management interventions, Hutton has noted that little information exists to assist with a reconstruction of the site's management history. In the 1920s a caretaker's cottage was built. The caretaker's wife used to serve refreshments to visitors to the reserve. Petrol rationing during the Second World War resulted in the temporary abandonment of the annual horse race meetings. The reserve continued to be managed by the Council of Newham and Woodend until 1959 when three members of the Hanging Rock Race Club were added to the Committee of Management. In 1968 a new cottage was erected and the caretaker's wife served refreshments from an old wooden shelter shed, now known as the hall, during the construction. In 1975 the Racing Club received funding from the Racecourse Development Fund to enable a large dam to be constructed in the centre of the racecourse to provide much needed water. In 1978 the Victorian Premier, the Hon. Rupert Hamer procured two grants of $40,000 each for the Management Committee, as well an interest free loan of $120,000. With these funds barbeques, toilets, and car parks were installed to cope with the increase in tourism that resulted from the release of the film 'Picnic at Hanging Rock' (see below).

> Through the years there have been countless picnics of every kind, large organised affairs for groups and societies as well as everyday outings for the family. The Reserve is still the chosen place for parties and celebrations, marriages and re-enactments and in recent years it has also been the site for activities on a grander scale – athletic carnivals, hot air balloons and vintage car rallies. There are one off events and no doubt like them will occur from time to time, but every day it is the tourists who come to experience and enjoy this intriguing rock formation. With great demands being made on it all the time, it is remarkable how well the small park has withstood the tremendous pressure of people during these past years (Hutton, 1991: 25).

In 1990 management of the reserve reverted to the local council. In January 1992 a visitor survey was conducted over a seven day period – questionnaires were handed to drivers of vehicles by the park ranger as the vehicles entered the reserve. A total of 238 useable questionnaires were obtained. The survey provided an indicative picture of the types of tourists who visited; their reasons for visiting; the nature of their visit; and the improvements they wanted to see. The survey revealed that most people visited the reserve as part of a day trip or they visited the site as part of an extended trip; most visitors spent less than half a day at the reserve. The weekend was the most popular time to visit the park; cars represented the most frequent mode of travel to the park. Some 10 per cent of visitors were international; and almost half came from Melbourne; and 20 per cent were from interstate. In terms of the activities that were pursued at the reserve; 83 per cent of respondents confirmed that they climbed the rock; 63 per cent walked around; 53 per cent picnicked; and 47 per cent relaxed. The primary reasons for visiting the reserve were the desire to climb and/or see the rock and visit the area made famous by the book and the film.

In 1993 consultants Loder & Bayly prepared a Management Plan of the reserve for the Committee of Management. The study coincided with the Committee of Management's expansion of the reserve when it purchased of an adjacent property of 22 hectares, today known as the east paddock. The management plan included three recreational goals: manage the use of the reserve to minimise impacts on conservation, historical and social values; maintain and enhance the diversity of recreational activities in the Reserve; and ensure that Reserve management does not detract from recreational amenity. Morgan & Lok (2000: 399) were critical of these objectives because they did not come with any specific, measurable or time bound determinants of acceptable social conditions. The lack of environmental conditions should also be noted.

In a study of the social carrying capacity of Hanging Rock, Morgan & Lok (2000) cited visitor statistics from 1996 that indicated that the Reserve received approximately 185,000 visitors each year, and that more than one-fifth of these were international visitors. More recently, the Victorian Department of Planning and Community Development (2013: 62) confirmed that Hanging Rock attracts over 120,000 visitors per annum, and is one of the most popular destinations in its region. Recent musical concerts at the site in the east paddock, including head-line acts such as Leonard Cohen, Rod Stewart, and Bruce Springsteen, have attracted crowds of over 61,000 at the events. The Rolling Stones are booked to perform at the Hanging Rock Reserve in late 2014. The Macedon Ranges Council had signed an exclusive agreement with the event company Frontier Touring which was up for renewal in 2013. A new five-year contract was signed off in October 2013. The contract allows the company to stage up to four concerts at the Rock each year. In June 2013 the Federal Government awarded $2 million through its Regional Development Australia fund to the Macedon Ranges Council to redevelop the Hanging Rock site. It was proposed to develop accommodation and conference facilities in the East Paddock and to improve walking tracks, signage, and public conveniences at the rock and to improve access for disabled visitors. Infrastructure such as the café and the interpretive centre also required upgrading. In August the Council approved plans to allow accommodation and a conference centre to be built at the Rock at the adjoining paddock added to the reserve in 1993; this immediately drew criticism from the Macedon Ranges Residents Association who was opposed to the commercial development.

6.4 Fourth Phase: Duplication

MacCannell's (1976) fourth phase in the development of attractions is that of 'duplication' or 'mechanical reproduction', when copies of the nucleus of the attraction, in this case the rock, are made available through media such as paintings, photographs, and postcards.

In 2012, the Gisborne and Mount Macedon Districts Historical Society published *Pictorial Hanging Rock a journey through time*. It is an extensive publication of images of Hanging Rock as represented by artists, geologists, photographers, and post cards that are in the society's collection at the Gisborne Court House.

In 1875, William Ford produced a large oil painting entitled 'Picnic at Hanging Rock, near Mount Macedon' (see Fig.6.6). Arnold Shore, writing in *The Argus* (27/5/1950) discussed the painting and its artist.

> One such, a man of whom very little is as yet known, except his name–William Ford–produced in 1875 a large oil painting entitled "Picnic Party at Hanging Rock, near Mount Macedon". This picture has recently been acquired by the Trustees of the National Gallery, on the recommendation of the director, Mr. Daryl Lindsay.[31]
>
> In dark tonality it portrays, quite notably, groups of women and children in the picturesque costume of the period–city visitors obviously contrasted with the strange setting of the rocks and trees of this old volcanic upheaval.
>
> All who have visited the Hanging Rock (known in early days as Mount Diogenes) must have felt something of the fascinating, rather "witches' kitchen" influence it imbues. The large masses of rock–the geological like of which exist, I am told, only in far-off Sweden–rear like monoliths, and disclose between their monumental forms vast panoramas of hill, mountain, and plains. Under the feet one senses one-time hollowness; but there is enough packed earth to permit the growth of proud eucalypts, which rear their erect or twisted forms in defiance of arboreal security.
>
> Much of the effect of all this has been surely observed and represented by William Ford. His picture, however, is in no way similar to those of our Australian Impressionist painters. Tonally, it might almost be a moon-light piece, it is so dark; but the eyes become accustomed to its darkness, and read a quite convincing effect of realism. This is what we are now learning-that there is no need for the artist to always seek to match the tonality of Nature. A great artist convinces us without matching endeavour.

In 1891, the Australian novelist Ivan Dexter published a novel entitled 'Mount Macedon Mystery', as a newspaper series of 26 chapters. The first chapter was entitled 'Macedon and Diogenes' (*Singleton Argus*, 8/8/1891). In the chapter he conflates Mt Diogenes with 'The Camel's Hump' and suggests that the appellation Diogenes was probably conferred 'on account of its rugged and forbidding character'. Dexter observes that there 'are many places in this mountain range that have never yet been trod by the foot of man, although they are surrounded by civilization, and so near a dense population'. 'Some of the narrow glens mentioned have probably no equal in natural or artificial loveliness, whilst portions of the range are gloomy beyond imagination, or terrible like the precipices and fastnesses of Mount Diogenes'. Dexter noted that although the peaks were reserved as a State Forest, he believed that in a 'comparatively few years as settlement advances this picturesque region will be shorn of its primeval beauties, for the hand of nature will have to give way to the hand of man'. This was the setting for his mystery for 'Many years ago a strange tragedy occurred

31 Sir Daryl Lindsay was director from 1942 until 1956. He is the husband of the author Joan Lindsay who wrote 'Picnic at Hanging Rock', which shared the same title as William Ford's 1875 painting.

in this district, the story of which will now be told the reader'. The story concerns a murder that occurred on Diogenes when a member of a party climbing the rock was deliberately pushed to his death, and his body buried in a wombat hole. The rest of the story is concerned with finding the man's remains and his murderer.

The most significant duplication that has helped to make Hanging Rock such a famous attraction is the fact that it was the setting of Joan Lindsay's 1967 fictional novel 'Picnic at Hanging Rock', which tells the story of a group of girls from Appleyard College visiting the rock on Valentine's Day 1900 for a picnic. Three school girls and later a school teacher go missing with one of the school girls being found later with no memory of what happened to her and her school mates. In 1975 the novel was the basis of an Australian movie directed by Peter Weir, also named 'Picnic at Hanging Rock'. The movie was set at Hanging Rock and rekindled interest in Lindsay's novel and saw a substantial increase in international and domestic tourists who visited the rock. The disappearance of the schoolgirls was explained in 1987 when the final chapter which had not been published with the Lindsay novel was published under the title 'The Secret of Hanging Rock'.

6.5 Fifth Phase: Social Reproduction

'The final stage of sight sacralization is social reproduction, as occurs when groups, cities, and regions begin to name themselves after famous attraction' (MacCannell 1976: 45). Social reproduction began with the establishment of the Hanging Rock Hotel in 1866. It is also possible to find the name in Hanging Rock Winery and Hanging Rock Cottage at Newham, and Hanging Rock House at Mt Macedon

6.6 Conclusion

Mt Diogenes, or Hanging Rock as it is more popularly known, was the subject of geological research by William von Blandowski in 1855; however it was not until Boxing Day 1864 that we have the first recorded account of the rock being used for leisure purposes when a picnic was organized by a group of tradesmen from Kyneton. In 1869 entrepreneur William Adams purchased the Rock and set about to develop the site into a pleasure resort. In 1866 he had purchased a half-acre block near the rock and had built the 'Hanging Rock Hotel'. In 1869 Adams placed a weir across Five Mile Creek to form a lake suitable for small rowing boats and other small water craft, and with the help of neighbouring publicans he developed a full programme of sports. Over the next seven years Adams cleared a race track for bush picnic races and built a pavilion for dancing and he continued to cater for picnics and sporting carnivals. In 1876 Adams sold the Rock to a Melbourne warehouse owner who had no interest in promoting a tourist attraction so the new owner leased the grounds to the owners of the Hanging Rock Hotel.

Around this time community agitation began to stir for the Government to purchase Hanging Rock for the public. The community of Newham urged its shire council to lobby the Government. In 1884, after a series of negotiations the Government purchased the Rock (and added it to the existing Crown Land reserve of 96 acres, west of Hanging Rock, which had never been privately owned), formally gazetting the Hanging Rock Recreation Reserve in the same year. The reserve was placed under the management of the Newham Shire Council. In 1885 a Racing Club was organized. The Council formed a committee of management from within its own councillors and oversaw site maintenance, pest control, and necessary site improvements. The races at Easter and New Year's Day were significant events, for example, in 1911 some 20,000 attended the News Years' Day races. In the 1920s a caretaker's cottage was built and refreshments were provided by the caretaker's wife to those who visited the reserve. In 1959 three members of the Hanging Rock Race Club were added to the Committee of Management.

In 1967 author Joan Lindsay published her novel 'Picnic at Hanging Rock' which was produced into a movie in 1975. The movie saw a substantial increase in international and domestic tourists who visited the Rock. In 1978 the Victorian Government made funds available for site improvements including barbeques, toilets, and car parks to cope with the increase of tourism that had resulted from the release of the film. In 1990 management of the reserve reverted to the local council. In 1993 the reserve was expanded with the purchase of an adjacent property of 22 ha. – it was now 88 ha in size.

In terms of the agency responsible for the emergence and development of tourism at Hanging Rock this case study differs from the others in this book for the early history of the Rock is tied up with the fact that it was alienated land and owned by an entrepreneur who could see the tourism possibilities the rock offered. From 1869 until 1884 the entrepreneur developed the site into a tourist resort. It returned to public land in 1884 only after community agitation.

References

Barned, B.J. (1985). *Woodend on the Five Mile Creek.* Woodend: The Author.

Blake, L.B.J. (1977). *Place Names of Victoria.* Adelaide: Rigby Ltd.

Blandowski, W. von. (n.d). *Terra Cognita : Foot of Diogenes monument.* Retrieved August 8,

Blandowski, W. (1855). Personal Observations made in an Excursion towards the Central Parts of Victoria, including Mount Macedon, McIvor, and Black Ranges' *Transactions of the Philosophical Society of Victoria, Melbourne,* 1: 50-74.

Butler, R.W. (1980). The Concept of a Tourist Area Cycle of Evolution: Implications For Management of Resources. *The Canadian Geographer, 24*(1), 5-12.

Clark, I.D. (1990). *Aboriginal Languages and Clans – an historical atlas of western and central Victoria, 1800-1900*. Clayton: Monash Publications in Geography, Monash University.
Department of Planning and Community Development. (2013). *Draft Loddon Mallee South Regional Growth Plan Background Report*. Melbourne: Department of Planning and Community Development.
Gisborne and Mount Macedon Districts Historical Society Inc. (2012). *Pictorial Hanging Rock, A Journey Through Time*. Mount Macedon: Gisbourne and Mount Macedon Districts Historical Society Inc.
Gunn, C.A. (1994). *Tourism Planning Basics, Concepts. Cases*. Washington: Taylor and Francis.
Heckenberg, K. (2011). Shifting terrain: vision and visual representation in Our Antipodes (1852) and Australia Terra Cognita (1855-6). *Journal of Australian Studies*, 35(3) 373-388.
Hutton, M. (1991). *The Hanging Rock*. Mount Macedon: Mt Macedon Historical Society.
Illustrated Australian News for Home Readers 25/2/1874
Loder & Bayly Consulting Group. (1993). *Hanging Rock Recreation Reserve Management Plan–a Report to the Committee of Management*. Melbourne: Loder & Bayly.
MacCannell, D. (1976). *The Tourist: A New Theory of The Leisure Class*. New York: Schocken Books.
Macedon Ranges Tourism. (n.d). *The History Of Hanging Rock*. Retrieved January 13, 2013, from Macedon Ranges Shire Council: http://www.visitmacedonranges.com/natural-attractions/hanging-rock/the-history-of-hanging-rock/
Mackay, M. (2011). Singularity and the Sublime in Australian Landscape Representation. *Literature & Aesthetics*, 8:113-127.
Morgan, D.J. & Lok, L. (2000). Assessment of a Comfort Indicator for Natural Tourist Attractions: the Case of Visitors to Hanging Rock, Victoria. *Journal of Sustainable Tourism*, 8(5): 393-409.
Poulter, J . (2011). *Sharing Heritage in Kulin Country–Lessons in Reconciliation from our First Contact History*. Templestowe: Red Hen Enterprises.
Singleton Argus 8/8/1891
The Age 2/1/1912
The Argus 28/12/1871; 17/10/1917; 27/5/1950
The Australasian Sketcher with Pen and Pencil 17/2/1877
The Illustrated Melbourne Post 25/1/1865

Fig. 6.1: William Blandowski, print after James Redaway & Sons, engraver
Diogenes Monument "Anneyelong" looking south towards Mount Macedon 1855-56
from the unpublished folio *Australia Terra Cognita*
engraving and aquatint, printed in black ink, from one copper plate
printed image 16.2 x 21.2 cm
plate-mark 21.2 x 28.2 cm
sheet 36 x 27 cm
National Gallery of Australia, Canberra
Gift of Major General T.F. Cape, in loving memory of his wife Elizabeth Rabett 1996

Fig. 6.2: William Blandowski, print after James Redaway & Sons, engraver
Foot of Diogenes Monument. 4 miles N. from Mount Macedon, 40 miles NNW from Melbourne 1855-56
from the unpublished folio *Australia Terra Cognita*
etching, engraving and lavis, printed in black ink, from one copper plate
printed image 16.8 x 21.6 cm
plate-mark 21.4 x 28 cm
sheet 36 x 27 cm.
National Gallery of Australia, Canberra.
Gift of Major General T.F. Cape, in loving memory of his wife Elizabeth Rabett 1996.

Fig. 6.3: 'Dryden's Rock, near Mount Macedon'. Robert Bruce, wood engraving in *The Illustrated Melbourne Post* 25/1/1865.

Fig. 6.4: The 'Hanging Rock': photograph Emily Slattery 2013

Fig. 6.5: 'Approach to the Hanging Rock, Mount Macedon' Source: *Illustrated Australian News for Home Readers* 23/4/1872.

Fig. 6.6: 'At the Hanging Rock 1875' William Ford, oil on canvas, National Gallery of Victoria

7 Mount Buninyong Scenic Reserve by Jaimie Watson and Ian D. Clark

This chapter provides a historical examination of Mount Buninyong, tracing its evolution from an Aboriginal cultural site into a recreational tourist attraction. It follows the example set by Clark's (2002) study of the evolution of Lal Lal Falls as a tourism attraction and documents the development of Mount Buninyong as a tourism attraction with significant consideration of key moments such as the first accounts of European visitation in 1837, its declaration as a public reserve in 1866, and the year 1926 when a road was built to the summit. To understand the history and evolution of tourism visitation to the Mount, this chapter will explain this transition by showing the significance of the works provided by MacCannell (1976; 1999), Butler (1980) and Gunn (1994). It will demonstrate that when these theoretical models are combined they contain explanatory materials capable of comprehending the history and evolution of a natural and cultural tourism attraction. MacCannell's studies into the development of secular attractions suggest they go through five stages including sight sacralization and/or naming, framing and elevation, enshrinement and duplication or mechanical reproduction, and finally social reproduction. Butler's tourism area life cycle model may explain the succeeding stagnation and rejuvenation of the Mount. Gunn's spatial model of three zones of visitor interaction should provide an understanding of the evolution and history of planning at Mount Buninyong.

Mount Buninyong is an extinct cinder cone volcano located 15 kilometres south east of Ballarat near the township of Buninyong. The summit of the mountain is between 719 and 745 metres above sea level and overlooks the City of Ballarat offering 360 degree views of the goldfields region (City of Ballarat, 2012). Mount Buninyong was created by an accumulation of cinder rocks that were ejected during explosive volcanic eruptions. Cinder cones generally range from about 50 to 200 meters in height, and consist of a steep sided mountain with a rounded top (Geocaching, 2008). Two types of rocks were produced from the volcano, cinder and basalt; both of which are hardened lava. The difference is scoria is blasted into the air which causes air bubbles to form making it light, while basalt is formed from lava flows that runs along the ground caused by the heaviness of the flow (Geocaching, 2008). Two separate volcanic cones exist at Mt Buninyong. The volcanic cone that is visible on site is the most recent, while the older volcanic cone has been destroyed mostly due to the eruption of the newer cones and erosions (Geocaching, 2008). The first volcanic cone at Mount Buninyong was created between 100 000 and 150 000 years ago, while the second cone occurred only 12,000 years ago (Geocaching, 2008). Mount Buninyong when it was an active volcano produced lava flows in two different directions; first to the East out of a flank crevice in the side and to the West over the summit of the volcano. These lava flows produced large lava plains at Lal Lal Falls (Clark, 2002). Mount Buninyong is an important refresh point for the groundwater system of the surrounding volcanic plains due to its highly permeable scoria cone. Springs are found at the base of the Mount where the more recent volcanic material rests on the underlying Ordovician frame (City of Ballarat, 1997).

Mount Buninyong is situated within the traditional country of the Wathawurrung Aboriginal people, and is one of Victoria's most significant geological sites. The area surrounding Mount Buninyong formed the estate of the Keyeet baluk, a sub-group of the Burrumbeet baluk clan, that was centred on lakes Burrumbeet and Learmonth (Clark, 1990). The Australian Heritage Commission has stated that it is possible that Aboriginal cultural values of national estate significance may exist for the Mount also (City of Ballarat 1997).

7.1 First Phase: Sight Sacralization and Naming 1837–1866

MacCannell (1976) has identified the first phase in the development of secular attractions as sight sacralization or naming. This takes place when a sight is marked off from similar objects worthy of preservation or the site has been given a name (MacCannell, 2001). Clark (2009) makes note that the naming of a place plays a vital role in the development of tourist attractions, and suggests 'If the names of sites are dysfunctional and fail as informative markers and conjure expectations that are not met, then names can become management problems and contribute to graffiti, vandalism and destruction and negative word of mouth publicity'. In relation to Gunn's (1994) spatial model of three zones of visitor interaction, the site is the 'nucleus' of the attraction; the main focus of tourist interest. In this case it would be Mount Buninyong itself. Butler's (1980) tourism area life cycle model can help to clarify these phases with his 'exploration' stage in which tourism is promising and visitor numbers are growing yet insignificant, a stage similarly described as 'pre-tourism'.

Buninyong is an Aboriginal toponym. There are variant spellings in the literature including Boning Yong; Bonningyong; Bun.un.yong; Bunin-youang; and Bunning-yowang. The earliest translation of the place name is provided by Katherine Kirkland (1845) who believed it meant 'big mountain'.

> We were now in the Boning Yong district, which takes its name from a very high mountain, on the top of which is a large hole filled with water. It is quite round, as if made by man, and there are fish and muscles [sic] in it. Boning Yong is a native name, and means big mountain. I like the native names very much: I think it is a great pity to change them for English ones, is as often done (Kirkland, 1845: 27).

Andrew Porteous (in Smyth, 1878 vol. 2: 179), who was the local guardian for the northern Wathawurrung in the 1860s and 1870s in the Carngham district, analysed the meaning of Buninyong as 'man lying on his back with knees raised'. William Withers (1887: 13), in his local history of Ballarat, claimed the name had the following origin:

> Buninyong, or, as the natives have it, Bunning-yowang, means a big hill like a knee – bunning meaning knee, and yowang meaning hill. This name was given by the natives to Mount Buninyong because the mount, when seen from a given point, resembled a man lying on his back with his knee drawn up.

Aldo Massola's (1962: 111) gloss was 'knee mountain', with links to the myth he learnt from Annie Alberts; the name means 'knee' since the mountain looks like an Ancestor lying down with his knee raised. Les Blake (1977: 182) translated it to mean 'man lying on back with raised knees'. The word for 'knee' in Wathawurrung is 'pun/bun' (Hercus, 1999). Barry Blake's analysis is that it means 'with his/her knees' (Clark & Heydon, 2002). Clark and Heydon's (2002) database of Victorian Aboriginal place names, confirmed that the name Bun-a-nyung also referred to Mt Fyans Home Station near Darlington.

In terms of Aboriginal mythologies about the creation of Mount Buninyong, William Stanbridge (1861) has published the following account he received from tribes in the neighbourhood of Fiery Creek. The account is interesting as it links two volcanoes together – mounts Elephant and Buninyong:

> One of the legends that these tribes are fond of relating is, that Tyrrinallum (Mount Elephant) and Bouningyoung (two volcanic hills about thirty miles apart) were formerly black men, that they quarrelled and fought, the former being armed with a leeowil and the latter with a hand spear, and after a prolonged contest, Tyrrinallum thrust his spear in Bouningyoung's side, the cause of the present hollow in the side of the hill, which so infuriated him that he dealt the other a tremendous blow, burying the point of the leeowil in his head, which made the present large crater, and knocked him to the spot where he now stands (Stanbridge, 1861: 300).

Massola (1962) recounted another version of this legend, told to him by Annie Alberts, who he considered to be the last full-blood Aborigine from the Western District. Albert's version of the legend is as follows (Massola, 1962: 110):

> Mount Elephant and Mount Buninyong were once men. Mount Elephant was in possession of a stone axe. Buninyong offered him some gold for it. Having agreed they met at what is now Pitfield Diggings for the exchange. Some time later Buninyong reconsidered, and desired his gold back. Elephant refused. Buninyong sent him a fighting message, and the challenge was accepted. They met at Pitfield Diggings. Elephant buried his spear in Buninyong's side, and the hole can be seen to this day. Elephant received a deadly blow on the head from Buninyong's stone axe. The gaping hole on Elephants head can also be seen to this day. The two men, mortally wounded, retired in opposite directions; their bodies turned into mountains, can be seen today at the spots where they died.

Massola (1962) comments that the newer version has 'post-European components', as gold was not valued in traditional times; it was only after the 1850s gold rushes that Aboriginal people began to value gold.

The first recorded European ascent of Mount Buninyong was in May 1837, when Frederick D'Arcy, a Government Surveyor, had learnt from the local Aborigines of the whereabouts of Mount Buninyong and Lal Lal Falls (Griffiths, 1988). D'Arcy, a surveyor, explored along the Moorabool River to Mount Buninyong. D'Arcy was accompanied by Dr Alexander Thomson and George Frederick Read, a prospective settler (Macqueen, 2010). George Frederick Read, who was accompanying D'Arcy on his expeditions and later become one of the earliest settlers of Buninyong, wrote the following journal account during May, 1837:

> I went with Dr. Thomson and D'Arcy the surveyor exploring the country as far as Buninyong from which hill we saw some large lakes in the distance and a country similar to that we had gone through the day previous which was certainly most beautiful. Rode down an emu on the way the flesh of which when fried was palateable... (quoted in Griffiths, 1988: 1).

In August, 1837 another expedition party was formed, again including D'Arcy and Dr. Thomson as well as George Russell, David Fisher, Captain Hutton, Thomas Learmonth and Henry Anderson (Macqueen, 2010). The party was led by an Aboriginal guide. They headed towards Lal Lal Falls and to the foot of Mount Buninyong with a cart and tent; Learmonth noted that it was 'the only hill that breaks the horizon to the north-west of Geelong' (Bride, 1898: 39). The party was unprepared with supplies to continue the expedition, and as a result most of the party camped on the site which later became Andrew Scott's 'Mt. Boninyong' homestead (Griffiths, 1988). Russell's account of this experience is as follows:

> We could see nothing of Mr D'Arcy's party with the dray, and went about in search of them, firing off guns and pistols until it got dark; but we received no reply in the way of gun-shots, or any other sign that they heard us; and as the night was dark we made up our minds to spend it where we were. We gathered a quantity of logs together and made a large fire, and sat chatting away to each other until we got sleepy. Then we lay down on the grass with our saddles for pillows, all feeling the want of something to eat, having had nothing to eat since morning. We spent the night the best way we could. Fortunately, although it was dark and cloudy it was mild (Macqueen, 2010: 108).

Macqueen (2010: 108-109) also records Russell's expedition climbing the mount:

> At day break some of the party, of whom I was one, went to the top of Mount Buninyong to get a view of the country. Some of the party, from being so long without food, didn't feel disposed for such a steep climb in the early morning.

Macqueen (2010: 109) questions whether their climb to the top of the mount was worthwhile, as he points out that as is the case today Mount Buninyong was covered in a 'healthy forest of eucalypts which allowed only glimpses of the landscape beyond'.

Andrew Scott established his run on the eastern side of the Mount in 1839. He named his run 'Boninyong' as it is still called today, held by the same family. The marshy creek area nearby became known as Scott's Marsh and later a community developed around the area that is now known as Scotsburn (DSE Heritage Database,

1998). The Chief Protector of Aborigines, George Augustus Robinson visited the Buninyong district in February and March 1840 and called on the Scotts. Robinson also acknowledged there was a 'dray road' which was used to fetch timber, he writes 'this will become the road of the Buninyong settlers'. Regarding Mount Buninyong, Robinson (Jnl 11/2/1840 in Clark, 2000) noted 'The top of and sides of the hill are thickly timbered so that it being difficult to get an extensive view and my horse, being unfit for riding, I declined ascending the top of the mountain'.

Springs at the base of the Mount were important sites for early settlement including three pastoral stations and the township of Buninyong (City of Ballarat, 1997). Thomas Learmonth estimated the Aboriginal population within a 48 kilometre radius from Buninyong was approximately three hundred people (Griffiths, 1988). Griffiths (1988: 4) makes note that the Aboriginal people of the Buninyong district did not have much interaction with the settlers but were 'often tempted to help themselves to grazing stock'. Only one fatality is documented. In April 1838, Terence McManus, a shepherd employed by Thomas and Somerville Learmonth, was killed at a place later named Murdering Valley (Griffiths, 1988). Subsequently, it did not take the Aboriginal people long to begin trading with the settlers. In the same month, George Frederick Read recorded in his journal, 'A great many natives came here today and exchanged skins for flour–they were extremely cunning' (quoted in Griffiths 1988: 4).

In the 1840s timber from Mount Buninyong was cut by pit sawyers and splitters and sent as far afield as Geelong (DSE Heritage Database, 1998). Trial Saw Mills Company soon replaced them. A large steam saw mill operation developed into a self-contained settlement at the eastern base of the Mount in the 1850s (City of Ballarat, 1997). The exposed east facade of Mount Buninyong remains as a legacy of this industry. In 1860 the Buninyong Borough successfully applied to secure the Mount as a Timber Reserve 'for mining and other purposes' (City of Ballarat, 1997: 10). Guidelines for timber harvesting were not clearly articulated or enforced, as a result within 30 years of European settlement Mount Buninyong was nearly stripped of trees (City of Ballarat, 1997).

The township of Buninyong was surveyed in 1848 by W. Smythe. Accordingly land sales first took place in May 1851 (DSE Heritage Database, 1998). Records predate the discovery of gold on 9 August 1851 by Thomas Hiscock, the Buninyong blacksmith. The discovery of the Ballarat goldfield was a consequence of Hiscock's find after he was searching for a stray cow (Beggs-Sunter, 2005). As a result, Buninyong developed and turned to the Mount as its symbol and icon, it was known as 'the busiest town in Victoria outside Melbourne and Geelong' (Beggs-Sunter, 2005) (see Fig.7.1). The seal of the first municipal council formed in 1859 depicted a rising sun behind the mountain (DSE Heritage Database, 1998).

Mount Buninyong's significance lies with its cultural and natural values, however as Clark (2002) makes note, similar to Lal Lal Falls, it is interesting that apart from providing an Indigenous name for the mountain, Aboriginal cultural values were not influential in the sacralization of Mount Buninyong. European settlers were more interested in the natural values of the site.

7.2 Second Phase: Framing and Elevation 1866–1884

The second phase of MacCannell's (1976) model of development of secular attractions is 'framing and elevation'. MacCannell refers to elevation as putting an attraction up for display, in the case of Mount Buninyong opening the site up for visitation. MacCannell subsections this phase into two types: protecting and enhancing (MacCannell, 2001). In relation to Mount Buninyong, this phase began in 1866 with the site being declared a Public Reserve and ended with the development of an unsealed road in 1884 leading up to the mount's summit (Leather, 2008). MacCannell's framing and elevation phase correlates with Gunn's 'inviolate belt' zone in his spatial model and in Butler's 'involvement' stage in his life-cycle sequence. Gunn's inviolate belt represents the area directly surrounding the nucleus of the site, Clark (2002: 5) describes it as 'the psychological setting for introducing the visitor to the attraction'. Mount Buninyong being declared a Public Reserve in 1866 ensured that it would stop uncertainties of de-forestation of the site or allow settlers to infringe on the access to the summit (DSE Heritage Database, 1998). This period can also be associated with Butler's 'involvement' stage, which is characterized by tourist visitations increasing and the formation of a significant tourism industry that develops around the attraction (Clark, 2002).

In 1866 Mount Buninyong was declared a Public Reserve, although as early as 1860 concerns were expressed about the de-forestation of the area. The reserve was for 247 acres after the Assistant Commissioner for Crown Lands wrote a report in 1865 advocating the need to place restrictions on timber harvesting because the removal of large trees had caused the understorey to disappear (DSE Heritage Database, 1998). Buninyong Municipal Council suggested that 'It is the duty of this Council to take steps as deemed expedient for the purpose of securing not only the whole of Mount Buninyong, but all the timbered land within three miles of the municipality' (DSE Heritage Database, 1998). Surrounding land was made available in 10 and 20 acre allotments to small landholders, who were usually dependent on a 'Common' to ensure their viability (City of Ballarat, 1997). A report in 1866 stated from the rush for 20 acre lots at Mounts Warrenheip and Buninyong 'care should be taken that the recreational areas of the mount be not alienated and the district be thus deprived of its two most health giving summer resorts' (Buninyong Municipal Council Minutes cited in DSE Heritage Database, 1998). For a number of years Mount Buninyong, as a Public Reserve, fulfilled this role. An early 1860s cottage still remains on the Geelong Road near Granny Whites Lane (City of Ballarat, 1997).

Edmund Leathes (1880: 35-9) has published an account of a visit to Mount Buninyong during a stay in Ballarat in the 1870s:

> The drives around Ballarat were delightful; the most enjoyable of all was that to Mount Buninyong. It was a drive of some miles to the village which was at the foot of the mount; there we left our buggy and walked to the summit. It was hard work, but, when we got to the top at last, we were well repaid for our exertions. The view of the surrounding country, which was well watered and wooded, was magnificent in beauty and extent, regarded from all points. Standing on a pinnacle, as it were, we looked north, south, east, and west. The mountain itself was beautifully wooded, and, had it not been for the dread of snakes, the birds and foliage – the latter interlaced with flowering parasites – would have made the descent alone enjoyable.

From 1874 the Committee of Management for the Reserve formalised the grazing function and advertised for grazing tenders. For the following 90 years, until the early 1960s, Mount Buninyong was constantly grazed by stock. Surrounding areas were burnt at low intensity every few years during much of the grazing lease period (City of Ballarat, 1997).

In January 1877, a fire occurred at Mount Buninyong, *The Argus* recorded the following information:

> A column of smoke was seen from Ballarat, arising from behind Mount Buninyong, between 9 and 10 o'clock last Saturday morning, looking, says the Star, as though the old volcano had once more awakened its died out fires. It could easily be seen, however, that the cause was a bush fire, and as the day grew the fire extended around the base and over the top of the mount, till at night the hill seemed covered with fire at almost every point seen from Ballarat. This prominent landmark in a blaze would have proved an interesting sight to onlookers had it not been for the thought that the farmers clustered around its base might lose by the fire the result of a year's labour. The fire was gazed at by wondering groups of people at every point in the streets of the town whence a good view could be obtained and the wooded sides of the mount continued to keep up their bright appearance for four or five hours after dark, when they began to pale, till, when morning broke, nothing but a few thin wreaths of smoke could be seen curling up from the lower portions of the hill. On Sunday night a few trees that remained a light on the top showed out with special brilliancy against the sky. A large number of persons visited the locality on Sunday to see what damage the seemingly terrible conflagration had caused, but we are glad to say that the fire was almost entirely confined to the Government reserves, including the mount, and no farms were much damaged. This was the result of the incessant labours of everybody living about the mount, who turned out one and all to keep the fire from touching the crops, which in some places adjoined the reserves. In their endeavours they were successful, the fire, as we have said, doing little damage besides burning up the under- wood and fallen timber covering three-fourths of the mount, a portion of that side facing Ballarat being comparatively untouched. On the other side, where the fire began, it burned from base to summit, and only stopped for lack of fuel, having burnt in some places right under the fence to the edge of the No. 2 Railway road. It was on the other side, however, that the greatest danger was to be feared, as there the greatest number of farms are situated (*The Argus*, 22/1/1877).

The cause of the fire was arsonist and farmer, Joseph Innes, who lived at the bottom of the Mount. On 20 January 1877 he deliberately lit six to seven grass fires around the Mount in an attempt to create a severe amount of damage, which he evidently did. Innes was arrested and jailed for his crimes (*The Argus*, 22/1/1877).

In the 1880s the population of Buninyong was declining, with a population of 1450 in comparison to the estimated population in 1872 at 2000 (Griffith, 1988). The *Telegraph* noted 'industry has dwindled down to a very small item. Farming, however, especially round the mount, has been going ahead' (*Telegraph* quoted in Griffith, 1988). In 1884, celebrations for the proposed extension of the railway line to Buninyong occurred. Construction of a carriage drive to the crater, zigzag path to the summit, and the summit of Mount Buninyong was levelled to provide picnic and camp sites (Leather, 2008).

7.3 Third Phase: Enshrinement 1884–1960

The third phase of sacralization is 'enshrinement' which MacCannell (1999: 45) refers to when the framing material that is used has itself entered the first stage of sacralization'. Clark (2002) makes note that the phase applies when there is an increase in the numbers of tourists and the attraction's reputation is increased. This phase is essential as it concerns the community services that are important to tourists (Gunn & Var, 2002). During this phase of enshrinement, the attraction developed its third spatial zone 'zones of closure'. As summarized by Gunn it is an outer vicinity of community influence of travel structures such as land uses for modern travel services–in the case of Mount Buninyong the road and the rotunda. As Clark (2002) mentions this enshrinement phase also correlates with Butler's 'development', 'consolidation' and 'stagnation' stages. These stages often happen very rapidly, tourism numbers significantly increase, and the tourism industry related to the destination dramatically changes. More than likely, the stagnation stage occurs represented by declining tourism numbers (Butler, 1980).

A significant period in regards to the enshrinement stage of Mount Buninyong was the tradition of New Year's Day picnics. An example of the local tradition of ascending the mount is provided in the following account from Josiah Hughes (1891: 147-8):

> On December 27th I went by rail to Buninyong, by appointment, where I met my cousin, Hugh Jones and his son, waiting for me, well stocked with Sandwiches and light refreshments, and equipped with a field-glass and stout walking sticks ready for the ascent of Mount Buninyong. Our way for a considerable distance was along the main road to Geelong, from which we turned to the left along rude country roads leading to the farm houses. In some places they were well-made and kept in repair, but as we went on we left these behind and came to wide road reserves, wide enough for a dozen cartways, in a perfect state of nature, trees and saplings affording excellent shade from the hot sun, and the cartway winding in and out among them. It was now midsummer, and what was deep mud in winter was now fine dust, in some places 3 to 5 inches deep. After traversing this for about two miles we began to ascend in earnest, and it is surprising what an amount of country is revealed to one's gaze in climbing this mountain, — unexpected valleys and hills of cultivated land, rugged precipices in unexpected corners; craters of extinct volcanoes, winding footpaths, giant trees, and fallen trunks of others, up to the very summit, which we reached by 12.45 noon. We were glad to rest ourselves with our backs against an enormous

> old tree, which lay there charred by the fires of many a pic-nic party. As we sat and refreshed ourselves with the good things provided, we were fanned by a deliciously cool breeze from the South, after which we got up and reconnoitred the surrounding landscape. The atmosphere was clear, and a great extent of country was visible, and with the assistance of our field-glass we could see to a great distance. As we sat at our refreshments, we faced Mount Warrenheip, about five miles off to N.N.E., with Mounts Hollow-back and Rowan and Spring Hill to the left at greater distances, the Black Hill, Mount Blackwood, Gordon's, Mount Egerton, and Three Sisters to the S.E., with a magnificent plain of agricultural land between, mostly settled and cultivated, dotted over with trees, fences, homesteads and roads, especially along the Moorabool Valley, and nearly all to the eastward of the old Geelong road, to the township of Clarendon (old Corduroy). Beyond this to S.W. the land appeared poor and unsettled. We removed to another position and looked direct south, to Geelong, 50 miles away, without any interruption, with the fertile Barrabool Hills a little to the right, Mount Hesse, Gellibrand and Pollock almost due south by a little west, and Mount Elephant and the great lake Corangamite and its tributary rivers to the south-west. The country intervening between us and these objects for at least 20 miles is principally occupied by squatting stations with scarcely anything to indicate human occupation, the land being poor. After this we removed our standpoints to the highest ground facing north and north-west, where we could trace in the varying distances the great dividing ranges of the country, which govern the course of the rivers. Those on the North running inland to join the great river Murray, and those on the South, being the Werribee, the Moorabool, the Barwon, the Hopkins, etc. run direct to the Southern Ocean. From here we could recognise the whole of Ballarat, ten miles away, as if lying at our feet, with all the surrounding townships named after the various diggings or creeks which created them; and more to the north and north-west we saw fine arable, cultivated land, interspersed with poppet heads and mullock heaps, and other evidences of deep mining operations, reaching from Bullarook forest, to Smythesdale, taking in Mounts Blowhard, Bolton, Emu, and Lakes Learmonth, Burrumbeet, and Wendouree. As a majestic back ground to the whole, the great dividing range as far as the Pyrenees and the Grampians. I am well aware that this description is more or less unintelligible to most people, and that it is impossible to realize it without some acquaintance with the country; but it is possible that some of my younger readers will have the opportunity, and to those who will ever visit Ballarat my advice is, ascend Mount Buninyong. To an old colonist of the early days, the names of these townships, diggings, mountains, and rivers are very familiar: and to me, who has been absent for 24 years, this survey from an elevation of 2500 feet above sea level was very pleasant, revealing the wonderful development which had taken place. ... We got back to Hugh Jones's house by four o'clock, and were glad to partake of the good things prepared for us.

This phase began with the establishment of a road to the summit in 1884 and included improvements to the road in 1926 when a low platform lookout was built also constructed on the summit (Leather, 2008). In 1932 the Bell Memorial Tower (an old mine poppet head) was erected and replaced the platform lookout. The picnic rotunda was also built around this time (Leather 2008). A simple tower was built in 1916, replaced by a 'poppet head' from Bendigo in 1928 which was later returned to its original home when a third tower was erected in 1979. C.P. Wilson an engineer from Buninyong reported to the Buninyong Shire Council on November 1, 1928 that the "poppet heads for the look-out on Mount Buninyong had been erected at a cost of £296, but that a further expenditure of about £100 must he incurred before it could be opened" (*The Argus*, 2/11/1928). Wilson noted that £275 had been received towards the cost, and that the Ballarat City Council had promised a second contribution of £25, and the further

£50 was expected from the Forest Commission (*The Argus*, 2/11/1928). The views are extensive from this high tourist vantage point, however this lookout was condemned unsafe and the present lookout was built in 1979, which still functions as a fire lookout tower (Thorpe & Akers 1982; DSE Heritage Database 1998). A concrete water tank for stock use was constructed in the crater area in the 1920s. Some of the land was cultivated for crops such as oats and potatoes. Ringlock fencing still remained in the Mount Buninyong Reserve from this period (City of Ballarat, 1997).

Mount Buninyong has a long history of passive recreation, perhaps best demonstrated by the succeeding look-out towers and the past tradition of New Year's Day picnics (City of Ballarat, 1997). In the 1920s the Buninyong Progress Association requested a ten year lease to build a lodge on the summit. An associated request for a 'good tourist road to the summit' was approved but the lodge was not (City of Ballarat, 1997: 11). In the 1930s the Progress Association took on board a 'beautification campaign' designed to soften Mount Buninyong with exotic trees and bulbs. The colours of the planting remnants can still be seen on the Mount today (City of Ballarat, 1997).

In December 1937, Mrs R.D. Nevett, formerly Divisional Commissioner of the Girl Guides' Association in Broken Hill, reflected on a recent visit to Mount Buninyong and her family's association with the district. She commented on 'vandalism' at the site:

> Flying between Mt. Warrenheip and Mt. Buningyong when approaching Ballarat, feeling of the deepest reverence surged through me as the sunlight caught the steel "look-out" erected on the summit of the latter mount as a memorial to my public-spirited father who, with other splendid pioneer citizens, made this lovely inland city with its main street named after Sturt, the explorer, and modelled on Princess-street, Edinburgh, with flower gardens, glorious trees, shrubs and statuary extending more than a mile down the centre. You can understand something of my pride after I had read in a record of my father's life sent to us by the Legislative Council of Victoria, of which he was a member, that one of his last acts in the House was to urge that sufficient money be granted to construct a road to the summit of Mt Buninyong, thus serving a dual purpose–the relief of unemployment and the addition of another attraction for the tourist.
>
> The grade of this one-way traffic road is a triumph for the engineer as it winds, very gradually round this mount, whose height is over 3000ft. Arriving at the summit one climbs the winding stairway to the "Look-out" and is rewarded by the splendid panorama that spreads on all sides. On clear days the waters of Corio Bay, Geelong, are discernible. This immense view is so fittingly symbolical of the great breadth of outlook and vision that my father undoubtedly possessed. When descending the stair way I was shocked to notice the work of wanton persons who had so thoughtlessly disfigured the steps and railings by cutting their names and initials thereon. Girl Guides let me give you an unwritten law-never defile any place by scratching or cutting your names anywhere. I have encountered this same vandalism at the Grampians, also at the Blue Mountains (*Barrier Miner*, 21/12/1937).

In January 1952 the shelter kiosk on the summit of Mount Buninyong was 'burnt down by vandals' (*The Argus*, 26/1/1952). In September 1954 the State Government announced that it was providing £940 for improvements to the Mount Buninyong and Lal Lal Falls reserves, providing the committees of management could find a one-to-four contribution. At Mount Buninyong the roadway around the mount was to be reconstructed, and tables and seats supplied in the reserve (*The Argus*, 27/9/1954). The mount was subject to vandalism in November 1955:

> VANDALS CAUSE HAVOC AT MT. BUNINYONG
>
> BALLARAT, Tuesday: Vandals have caused considerable damage to installations at Mt. Buninyong summit and Mt. Buninyong and Lal Lal reserves. Vandals are also blamed for the serious fire which swept through the area recently. Repeated acts of destruction have the reserves committee worried, chairman Mr. P. E. Selwyn Scott said today. In a tour of the area to day Mr. Scott pointed out damage caused by vandals. Hundreds of beautiful messmate trees are still blackened and bear the scars of the fire. Mr. Scott said one of the most serious effects of the fires was the killing of regrowth of timber. One side of the Mount was becoming bare. The fire in February destroyed the shingle roof of the rotunda on the summit. Vandals also unscrewed the tap of a 5,000-gallon water tank. They let all the water out, and took the tap. They also took the downpipe from the roof into the tank. A rock fireplace was smashed and an attempt made to cut down a guide post indicating the down route for motor traffic from the summit. Absence of this sign could result in a serious accident to motorists trying to descend the mountain by the wrong route and meeting the up traffic. The Cr. Alex Bell memorial sign has been wrenched from the side of the staircase to the tower. The temporary shelter for fire spotters has been thrown down. Most serious damage was the removal of a large bronze direction tablet on the top of the 75ft. high tower. It was broken into three pieces. Removal of this plate must have taken considerable effort, and it is thought a hacksaw was used. The safety fence around the platform at the top of the tower was pulled loose at two points (*The Argus*, 16/11/1955).

In the 1960s the road to the summit was sealed during a period of time when grazing leases were terminated, but at this time the Buninyong Shire Council permitted a telecommunications tower to be constructed on top of the Mount (DSE Heritage Database, 1998). The multifaceted tower is large and fenced with hurricane steel mesh and the buildings are of apricot brick, with many large satellite dishes (DSE Heritage Database, 1998). In 1985 a further meshed area of much smaller scale was permitted and contains a tower for police communication. In the late 1990s there was an application for a further tower to be built (DSE Heritage Database, 1998).

7.4 Fourth Phase: Duplication

'Duplication' is MacCannell's fourth phase in the development of attractions when copies of the nucleus of the attraction, in this case the mountain, are made available through media such as paintings, photographs, and postcards.This phase is also known as 'mechanical reproduction' (MacCannell, 1999). The nucleus of the attraction is reproduced, in regards to Mount Buninyong duplicates are made through paintings and photographs.

The Mount has been associated with artist S.T. Gill who titled his sketch in the mid-1850s 'Ballarat from Mt. Buninyong' as well as Henry Winkles who in 1852 made a lithograph of the township of Buninyong which included the Mount in the foreground (Thorpe & Akers, 1988; Faulkner, n.d) (see Figs.7.2 & 7.3 & 7.4). Archibald Vincent Smith was a photographer who also took various photographs of the Mount in 1866 and Ballarat district (Trove, 2013).

7.5 Fifth Phase: Social Reproduction

For MacCannell, the final stage of sight sacralization is social reproduction which 'occurs when groups, cities, and regions begin to name themselves after famous attractions'. In the case of Mount Buninyong, social reproduction has occurred in the naming of the Boninyong pastoral run, Buninyong township, Buninyong primary school, and businesses such as Mount Buninyong Winery. Many businesses in Buninyong have the name Buninyong as part of their business names.

7.6 Tourism at Mount Buninyong

In the 19th century, Mount Buninyong was a very popular scenic attraction. Walks ambled from the foothills up through the crater to the summit with a simple wooden lookout platform. Later additions to the attraction included a rotunda-styled shelter, toilets and barbecue facilities (Leather 2008). Succeeding generations climbed the Mount unaided by tracks and picnicked in the crater–notably on New Year's Day (DSE Heritage Database, 1998).

The Reserve was recommended as a Scenic Reserve by the Land Conservation Council in 1982 (City of Ballarat, 1997). It is a popular destination for a drive out of Ballarat and is listed on local tourist brochures. It also offers a popular running and jogging track for fitness devotees (DSE Heritage Database, 1998). Mount Buninyong is much used by tourists, locals and visitors who enjoy the views from the tower of the summit and a walk in the beautiful surroundings. Snowfalls are a feature in winter. There are walking tracks and a road to the summit and down the other side. This area is a significant link with the Union Jack Reserve, north of Buninyong township, which is clothed in native scrub after mining activity in the 19th century (DSE Heritage Database, 1998).

Mount Buninyong Scenic Reserve offers an opportunity to experience a different bush land setting than is generally found in the region (City of Ballarat, 1997). The reserve has significant natural values. Its scoria cone carries an exclusive vegetation type of regional and state importance as one of the few scoria cones in Western Victoria which has retained native vegetation and provides spectacular views of the surrounding landscape (City of Ballarat, 1997).

In 1980 the present Mount Buninyong tower was opened to the attraction and in 1982 an information board and rotunda was constructed, and water tower was painted. Picnic seats and barbeques were put in place as were the formation of nature tracks and a new car park area was established (Leather, 2008).

In 1997, the City of Ballarat directed major recommendations for the Mount Buninyong Scenic Reserve including:

Biological surveys will be conducted to identify all significant flora and fauna.

- A vegetation management program will be introduced to restore understory species and to protect significant species and communities.
- Monitoring programs will be established to evaluate vegetation management techniques.
- Linkages will be developed from the reserve to nearby remnant vegetation.
- Fire management will be carried out using a permutation of slashing and the development of an ecological burning program for the reserve.
- New visitor facilities will be developed at the entrance to the Reserve and will include orientation information for the Reserve and its adjoining linkages. Existing visitor facilities will be rationalised.
- The existing walking track system will be upgraded and improved to enable return loop walking opportunities.
- The Mount Buninyong Access Road will be managed to ensure visitor safety
- Visitor enjoyment will be enhanced by the provision of interpretation of the Reserve's natural and cultural features as well as those of the surrounding region.
- Community awareness and involvement.
- Investigate and document Aboriginal history and protect identified sites as appropriate, in consultation with the local Aboriginal community.
- Involve the local Aboriginal community in developing interpretation for the Reserve and environs which relates to Aboriginal cultural heritage.
- Remove ring lock fencing from the reserve.

7.7 Conclusion

As a tourist attraction, Mount Buninyong–with its natural and cultural values, has transitioned from being an attraction for farming purposes in colonial days and a scenic attraction for picnics to being a highly recognizable public reserve (City of Ballarat, 1998). This paper has attempted to explain this transition by showing the significance of the works provided by MacCannell, Butler and Gunn. This paper demonstrated that when these works are combined they contain explanatory materials in comprehending the history and evolution of a natural and cultural tourism attraction.

Mount Buninyong is an important site for the acknowledgment of the ongoing connection of Aboriginal people with the land through creation stories and cultural sites and is an ideal site for interpreting Aboriginal cultural history (City of Ballarat, 1997). The Australian Heritage Commission has stated that it is possible that Aboriginal cultural values of national estate significance may exist for the Mount. This statement of significance is for a place with natural significance from a scientific and interpretive perspective as well as from a geological and geomorphic perspective. For this place the cultural values are more significant and are historic as well as social.

Mount Buninyong is an area of high geological and geomorphological significance. It is an excellent example of an amalgamated volcano; that is, as the Department of the Environment and Heritage (2003) describes as one which has both a breached scoria cone and associated lava flows. The place also provides a good illustration of the varied topography that has developed on adjacent lava flows of differing ages. In addition, it is a good example of the effects of aspect on plant communities (Department of the Environment and Heritage, 2003). Mount Buninyong has excellent interpretive value as it is the most obvious example in Victoria of a breached scoria cone and associated lava flows. The site is easily accessible as it is the only major scoria cone with a deep crater that remains on public land in Victoria (Department of the Environment and Heritage, 2003). The Mount is still used as a scenic attraction today, the bond between the local community and Mount Buninyong has always been strong. This has been expressed through organized interest groups as well as in the varied recreational pursuits of individual residents. In relation to Butler's life cycle model, the Mount could be said to be at the 'rejuvenation' stage as it is continuously being maintained and upgraded. In early 2012, the *Ballarat Courier* undertook a survey of the Ballarat region's favourite natural attraction. Lal Lal Falls was considered the region's best, securing 37.5 per cent of the vote; Lake Wendouree came in second with 32.3 per cent, and Mt. Buninyong, third, with 9.4 per cent (*Ballarat Courier*, 8/2/2012).

References

Australia. Department of Environment and Heritage. (2003) *Protecting National Heritage* (2nd ed). Canberra: Australian Heritage Commission.

Australia. Department of Sustainability, Environment, Water, Population and Communities. (1998) *Australian Heritage Database*. Canberra, Australia. Retrieved 3 November 2012 from http://www.environment.gov.au/cgibin/ahdb/search.pl?mode=place_detail;place_id=100370

Ballarat Courier 8/2/2012

Barrier Miner 21/12/1937

Beggs-Sunter, A. (2005). *Eureka–The Buninyong Connection*. Buninyong: Ballarat Reform League. Retrieved 10 December 2012 from http://www.ballaratreformleague.org.au/buninyong.htm

Blake, L.B.J. (1977). *Place Names of Victoria*. Adelaide: Rigby Ltd.

Bride, T.F. (Ed.) (1898). *Letters from Victorian pioneers: being a series of papers on the early occupation of the colony, the aborigines, etc.* Melbourne: RS Brain.

Butler, R.W. (1980). The Concept of a Tourist Area Cycle of Evolution: Implications for Management of Resources. *Canadian Geographer*, 24:5-12.

City of Ballarat. (1997). Centre for Environmental Management. *Mount Buninyong Scenic Reserve Final Management Plan*. Retrieved 12 December 2012 from *www.ballarat.vic.gov.au/.../mt%20buninyong%20scenic%20reserve*

City of Ballarat. (2012). Mt Buninyong Scenic Reserve. Retrieved from http://www.ballarat.vic.gov.au/lakes-parks-recreation/nature-reserves/mt-buninyong-scenic-reserve.aspx

Clark, I.D. (ed.) (2000). *The Journals of George Augustus Robinson, Chief Protector, Port Phillip Aboriginal Protectorate, Volume One: 1 January 1839-30 September 1840*. Clarendon: Heritage Matters.

Clark, I.D. (2002). The ebb and flow of tourism at Lal Lal Falls, Victoria, a tourism history of a sacred Aboriginal site. *Australian Aboriginal Studies*, 2:45-53.

Clark, I.D. (2009). Naming Sites: Names as management tools in indigenous tourism sites -An Australian Case Study. *Tourism Management*, 30:109-111.

Clark, I.D. & Heydon, T.G. (2002). *Database of Aboriginal Placenames of Victoria*. Melbourne: Victorian Aboriginal Corporation for Languages.

Faulkner, S. (n.d.) Brim Brim Gardens. Retrieved 12 December 2012 from http://www.brimbrim-gardens.com.au/history.htm#bun

Geocaching. (2008). Mount Buninyong Crater. Retrieved 12 December 2012 from http://www.geocaching.com/seek/cache_details.aspx?guid=e255a4e8-facf-4624-b9f1-23c7ed523074

Griffiths, P.M. (1988). *Three Times Blest: A History of Buninyong and District 1837-1901*. Buninyong: Buninyong & District Historical Society.

Gunn, C.A. (1994). *Tourism Planning Basics, Concepts, Cases*. Washington: Taylor & Francis.

Gunn, C., & Var, T. (2002) *Tourism Planning Basics, Concepts, Cases*. New York: Taylor & Francis.

Hercus, L.A. (1999). Place Names Notes and Papers relating to the Place Names Dictionary Project, Victorian Aboriginal Corporation for Languages, Melbourne.

Hughes, J. (1891). *Australia Revisited in 1890, and Excursions in Egypt, Tasmania, and New Zealand. Being extracts from the diary of a trip around the world, including original observations on colonial subjects and statistical information on pastoral, agricultural, horticultural, and mining industries of the colonies*. London: Simpkin, Marshall, Hamilton, Kent & Co.

Kirkland, K. (1845). Life in the Bush. By a Lady. *Chambers's miscellany of useful and entertaining tracts*, William and Robert Chambers, London, 1(8):1-39.

Leather, D. (2008). *Buninyong: Mount Buninyong*. Retrieved 11 December 2012 from http://www.buninyong.vic.au/mount.htm

Leathes, E.D. (1880). *An Actor Abroad: Or, Gossip Dramatic, Narrative and Descriptive, from the Recollections of an Actor in Australia, New Zealand, The Sandwich Islands, California, Nevada, Central America, and New York*. London: Hurst and Blackett.

MacCannell, D. (1976). *The Tourist: A New Theory of the Leisure Class*. New York: Schoken.

MacCannell, D. (1999). *The Tourist: A New Theory of the Leisure Class* (Revised edition). California: University of California Press.

MacCannell, D. (2001). Sight Seeing and Social Structure. In S.L. Roberson (Ed.), *Defining Travel: Diverse Visions* (pp. 13-29). Mississippi: University Press of Mississippi.

Macqueen, A. (2010). *Frederick Robert D'Arcy: colonial surveyor, explorer and artist c1809-1875*. N.S.W: The Author.

Massola, A. (1962). Two Aboriginal Legends of the Ballarat District. *Victorian Naturalist*, 79(4):110-111.

Smyth, R.B. (1878). *The Aborigines of Victoria*, 2 vols. Melbourne: Victorian Government Printer.

Stanbridge, W.E. (1861). Some Particulars of the General Characteristics, Astronomy, and Mythology of the Tribes in the Central Part of Victoria, South Australia. *Transactions of the Ethnological Society of London*, 1:286-304.

The Argus 22/1/1877; 2/11/1928; 26/1/1952; 27/9/1954; 16/11/1955.

Thorpe, M.W. & Akers, M. (1982). *An Illustrated History of Buninyong*. Buninyong: Buninyong & District Historical Society.

Trove. (2013) The Crater of Mount Buninyong. Retrieved 13 January 2013 from http://trove.nla.gov.au/work/12516621?q=creator%3A%22Smith%2C+Archibald+Vincent%2C+1837-1874%2C+%28photographer.%29%22&c=picture&versionId=14787987

Withers, W.B. (1887). *The history of Ballarat, from the first pastoral settlement to the present time* (2nd ed.). Ballarat: F.W. Niven & Co.

Fig. 7.1: Untitled drawing showing Mt Buninyong by J.B. Henderson, c. 1853-1856. State Library of Victoria Pictures Collection, Accession No. H28122.

Depicts two lines of huts at the base of Mt. Buninyong.

Fig. 7.2: The township of Buninyong, Victoria drawn & engraved by H. Winkles. Engraving 1855. (In private collection of Ian Clark)

Fig. 7.3: 'Buninyong Hill'. Wood engraving *The Illustrated London News* 29 May 1852.

Fig. 7.4: 'Mount Buninyong, Near Ballaarat'. Wood engraving by George Stafford 1861, reproduced in *The Illustrated Melbourne Post*, April 1862, George Slater, Melbourne. State Library of Victoria Pictures Collection.

8 Tower Hill State Game Reserve by Lisa Justin and Ian D. Clark

This chapter offers a historical analysis of the evolution of tourism at the Tower Hill State Game Reserve. Tower Hill is an extinct volcano and a major natural landmark located on the Princes Highway between Warrnambool and Port Fairy in South West Victoria. Today, the reserve is an important focus for tourism and protects a variety of environmental, landscape and cultural values. It provides visitors with walking tracks, picnic facilities, lookouts and the Worn Gundidj Visitor Centre is a venue for environmental and cultural interpretation.

A nested caldera, Tower Hill is one of the most recently active volcanoes in western Victoria (Downes, 1961:3). It measures 3.2 km by 2.4 km wide and is the largest of its type in Australia. The volcanic formation of Tower Hill comprises of a main crater with a surrounding tuff ring. The inner walls of the tuff ring are steep and enclose a flat-floored crater over 90 metres deep. Within the crater Tower Hill is divided into three scoria islands, and a permanent and semi-permanent lake. The reserve comprises 610 ha that are made up of three islands: Main Island (125 ha); Fairy Island (18 ha); and Hat Island (1 ha), and Tower Hill Bank (120 ha), and permanent and semi-permanent water.

Parks Victoria (2012: 1) stated that 'the way in which the volcano was formed is of great interest to geologists'. The caldera was created 30,000 BC after a violent volcanic explosion, which created a funnel shaped crater (which later filled with water to create a lake), islands and a volcanic cone.

> A change in the nature of the eruption led to the growth within the main crater of numerous small cones, built up of scoria and coarse volcanic fragments. These small cones are described as 'nesting' within the main maar crater. Several craters can still be seen in the small cones (Fisheries and Wildlife Department, 1983: 4)

It lies at the end of a line of thirty similar volcanoes, which extends from Colac to Port Fairy, and is part of the UNESCO-endorsed Kanawinka Geopark (Parks Victoria, 2012: 67).

The first confirmed European sighting of Tower Hill was by French explorers sailing with Captain Nicolas Baudin aboard the 'Geographe' on 31 March 1802. Francois Peron (1809: 245) has provided us with the earliest description:

> As we continued our course along the shore we came ahead of a peak of a conic form, to which we gave the name of the Peak of Reconnaissance, and the nearest cape we called Tabor Cape.[32]

32 Downes (1961: 11) has identified Tabor Cape as the headland at present-day Warrnambool. Baudin's (in Learmonth 1934) account does not mention the Peak of Reconnaissance.

Matthew Flinders also sailed along the southern coast of Australia and on 20 April 1802 in the ship's log referred to Tower Hill as 'Peaked hill, position uncertain'.

Tower Hill is situated within the traditional country of the Koroitgundidj Aboriginal people, who spoke the Peekwurrung dialect of the Dhauwurdwurrung language (Clark, 1990). Food sources were particularly plentiful in the fertile coastal area around Port Fairy and Tower Hill, with the western plains supporting the Aborigines' rich cultural life. For Aboriginal people the land was an intrinsic part of cultural and spiritual life, with natural features representing deep religious or 'dreaming' significance. Archaeological surveys of the area have uncovered axe heads and other artefacts in the volcanic ash layers indicating that there was an Aboriginal presence in the area before the eruption (Parks Victoria, N.D).

H.F. Wickham corresponded on the matter of Aboriginal people and volcanism in a letter to *The Sydney Morning Herald* on Friday 13 June 1902:

> On this subject [of volcanic eruptions] Mr James Dawson says -"An intelligent aboriginal distinctly remembers his grandfather speaking of fire coming out of Bo'ok, a hill near the town of Mortlake, Victoria when he was a young man " Another says -"When some of the volcanic bombs found among the scoria; at the foot of Mount Leura, were shown to a Colac native, he said they were like the stones which his forefathers told him were thrown out of the hill by the action of fire." Again, at Tower Hill, in the same State, beneath a bed of lava, "tuff,' the skeleton of a dingo was found buried, now as the dingo is supposed to have accompanied man to Australia, and man is the most recent development of the countless stages in the great scheme of evolution, it follows that the eruption which overwhelmed the dingo is of modern date.

Geologist Edmund Gill (1938) has also discussed Aboriginal associations with volcanism in the region. He noted that the Aboriginal people had a tradition that their ancestors saw 'burning mountains' in the Western District:

> Their description of a volcano as a "burning mountain" reminds one of the old description we were given at school "a volcano is a mountain which emits fire and smoke." The writer once heard a noted professor of geology comment upon that definition as follows:-"The definition is quite correct except that a volcano is not a mountain and it emits neither fire nor smoke. A volcano, strictly speaking, is a vent. It may exude molten lava, but it does not burn. The cloud of steam rising from the crater is often mistaken for smoke.
>
> However, are we to give any credence to the aboriginal legend? Has Tower Hill been active since the district was inhabited by homo sapiens? Can the volcano's life history be written with any accuracy? The geologist, hammer a-hand, has chipped out a great deal of that story. Detective-like, he has followed the clues of broken strata and yawning craters. He has examined the "finger-prints" of volcanic ash over the environing countryside. He has "reconstructed the crime," so to speak. And this is the dramatic story which he has spelt out from those rocks, a story which does not deny the aborigine his legend. ... Anyway, the Americans have a replica of Tower Hill at Crater Lake, in Oregon. That volcano also has been gutted. Like Tower Hill also it has built for itself a miniature mountain in the centre of the big truncated crater, and made that mountain an island by surrounding it with water. In recent years Tower Hill has lost most of its lake. It is reported that the tapping of spring waters for the supply of the nearby township of Koroit is the

explanation of this diminution. The aerial photograph shows part of the sides of the original mountain. At the foot of this rim are the flats, which until recently were covered with water. In the centre are the younger cones and craters–the dying work of the volcano. The remaining area of lake is just out of the photograph to the right. In the fore ground are some of the very fertile fields near Koroit.

But to come back to the aborigines' legend. How is that substantiated? For this part of the story we need to go five or six miles away to the city of Warrnambool. This attractive city is built on dune sandstone, of which many of its buildings are constructed. Some years ago there was taken from this sandstone the impression of an aborigine and his lubra where they had been seated in the sand. In another place probable footprints were discovered. Chance was kind on these occasions and preserved these impressions for us. Now the ash from Tower Hill overlies this dune sandstone; therefore the volcano must be younger than these human impressions. Tower Hill must have been active after human beings came to live in that district. Further evidence is provided in the fact that the strata of volcanic ash at Koroit overlay a modern sea beach strewn with shells identical in species and condition with those now found thrown up in the vicinity at high tide. To this extent, then, the scientist can corroborate the ancient tradition of the aborigine (*The Argus*, 18/6/1938).

The Aboriginal association with volcanism is also confirmed by their name for Tower Hill (see below) which is suggestive of eruptive activity.

8.1 First Phase: Sight Sacralization and Naming 1802–1891

In terms of MacCannell's (1976) first phase in the development of attractions – that of sight sacralization or naming, numerous Aboriginal names have been recorded for the caldera which was known as Koroit and the many features surrounding it. The Dhauwurdwurrung or Gundidjmara Aboriginal clan associated with Tower Hill was known as the Koroitgundidj (meaning 'belonging to Koroit') (Dawson, 1881; Clark, 1990). Clan names were typically taken from named places, so the clan name indicates the Aboriginal name of Tower Hill. They were also known as the Kartgundidj (meaning 'belonging to mountain') which is a descriptive name akin to 'mountain people' and was probably conferred on them by other clans.

Aboriginal microtoponymy at Tower Hill has been examined by Clark (2009) in which he identified that some eleven place names exist at Tower Hill, for features as diverse as scrub, flats, craters, the crater's lake, and lake banks (See Table 8.1). All eleven names are sourced from the ethnography of James Dawson (1881), which was the result of extensive study of western Victorian Aboriginal languages by Dawson and his daughter Isabella, from 1844.

Tab. 8.1: Aboriginal toponyms at Tower Hill

Aboriginal Name	Meaning
Tower Hill; *Koroitj*	Possibly nettles (Dawson, 1870), or suggestive of volcanism (Mulder, 1909: 11); note, local clan is Koroitgundidj (Robinson Jnl, 24/3/1842 in Clark 2000b).
Tower Hill – scrub; *Murrheaal*	Scrub between Tower Hill Flat and lake (Dawson, 1881: ixxxi).
Tower Hill Flat; *Taerra Mukkar*	Sweet root like a parsnip; West side of Tower Hill Flat (Dawson, 1881: lxxxii).
Tower Hill Flat – scrub; *Yaal*	Scrub on West side of Tower Hill Flat (Dawson, 1881: lxxxi).
Tower Hill Island; Parrang Kuttcha	Name of an edible root found there; The island surrounding by Tower Hill Lake, once joined with the sea (Dawson, 1881: lxxxiv).
Tower Hill Island – crater; *Yatt mirng*	White eye; crater in Tower Hill Island (Dawson, 1881: lxxx).
Tower Hill Lake; *Kurruk baruum*	Grandmother of lice; outlet from Tower Hill lake (Dawson, 1881: lxxx).
Tower Hill Lake; *Mitjil*	Mitj = skin (Dawson, 1881; lxxx).
Tower Hill Lake – bank; *Koroitj*	West bank of Tower Hill Lake (Dawson, 1881: lxxx).
Tower Hill Lake – bank: *Mum ngamat*	Bottom of the sea; East bank of Tower Hill Lake (Dawson, 1881: lxxxi).
Tower Hill Lake – outlet; *Kuuro baruum*	Grandmother of lice; Outlet of Tower Hill Lake (Dawson, 1881: lxxx).

The Dhauwurdwurrung Aboriginal name for Tower Hill is Koroit which survives in the nearby township of the same name. According to Dawson (1870), the name is suggestive of volcanic activity; and Mulder (1909: 11) has suggested it means 'smoking, hot ground'. As well as the archaeological record, toponymic analysis confirms Aboriginal presence during the last eruption (Downes, 1961: 5).

The first European name for Tower Hill was 'Peak of Reconnaissance' conferred in 1802 by the French exploration expedition led by Nicolas Baudin. The expedition included the cartographer-surveyor Louis-Claude De Freycinet and the naturalist Francois Peron. A few months later, Matthew Flinders' exploration party simply named the caldera 'Peaked Hill'. Given European discovery and naming of Tower Hill in 1802 ('Peak of Reconnaissance'–a name that did not survive in local parlance), Downes has suggested that Tower Hill would have served as a landmark for ships entering Port Fairy and Warrnambool, and although there was much maritime activity along the southwest coast over the next thirty years, few records survive and no direct references to Tower Hill by these seamen have been found. Downes posits the first examination of Tower Hill by land must have been by men employed as sealers and whalers along the southwest coast. He believes it probably first became known to land travellers in 1839 (Downes, 1961: 16).

Downes believes the popular name 'Tower Hill' came into use in the early 1840s, although when he conducted his research the earliest formal use of this name that he could find was Pickering (1846). In April 1841, the Chief Protector of Aborigines in the Port Phillip District, George Augustus Robinson, visited the Port Fairy district, and his private journal reveals that the eminence on the edge of the swamp was called by 'Campbell's men "Tower Hill"'. Campbell is a reference to Captain Alexander Campbell, who was in partnership with John Griffiths and Michael Connolly in an agricultural enterprise on a stretch of land extending from Killarney to the Hopkins River from 1838 until 1842. Robinson's reference provides us with a date of usage of 1841 (and possibly earlier), and infers that it was the name used by station employees, and possibly conferred by them. Downes considers that Tower Hill is a descriptive name, and posits that the volcanic formation may have resembled a castle complete with a moat and steep walls, hence the name. Anita Brady (1992: 2) has offered a second explanation: 'Another account credits Dougald Langtree, a sailor from Glasgow with naming the site after Tower Hill in Scotland'. Presumably Brady is referring to Tower Hill, overlooking Gourock on the Clyde coast.

When Robinson learned of Tower Hill on 28 April 1841 he also learned that a 'native village' was to be found there. He learned that 'There is a large swamp on the east of the Port Fairy river where the natives get their chief support, roots, &c., and near to a small eminence on the edge of this swamp, called by Campbell's men 'Tower Hill', is a native village: an assemblage of huts &c. Along the coast near to the shepherds huts are reefs of rock abounding with the haliotus mutton fish, muscle [sic], moreover crawfish and which must have afforded them a good supply as the camping place of the natives in the beautiful forest of dwarf banksia abound with cinirated shells' (Robinson Jnl 28/4/1841 in Clark, 2000a).

The area around Tower Hill was largely ignored by squatters because its rich volcanic soil supported dense vegetation. They were looking for open grazing land and the land around Tower Hill was too thickly forested (McCorkell & Yule, 1999). Between August and December 1840, Tower Hill was incorporated into an extensive pastoral run held by the Bolden brothers (Armyne, George, and Lemuel). In November 1841 William Rutledge purchased part of this run for a special survey. In 1843 the Ryrie brothers (Alex and William) took over the Lake station from the Bolden brothers, and Tower Hill was its central feature. In 1844 Jack Buchanan was listed as the license holder. Rutledge is listed as the station holder from 1849 until 1851.

Richard Hoddle, who carried out the first parish survey in 1845, described the area as having chocolate covered soil of the richest description (McCorkell & Yule, 1999). Reports such as this brought the earliest pioneers to the district to clear the forest and establish farms. The soil from Tower Hill determined the pattern of settlement and the development of Koroit (McCorkell & Yule 1999). Little was recorded of Tower Hill until a map was compiled from surveys done between 1846 and 1850. The surveyors confirmed that the land rising toward the banks of Tower Hill was heavily timbered with thick forests of stunted gum and a heavy undergrowth of ferns and native laburnum.

In the 1850s, the Midgley family settled at Yangery Creek near Tower Hill and lived in tents until they hacked out of thick bush a suitable plot to build their home. Sarah Midgley described the view from their tents as 'that desolate spot surrounding with dense scrub, stringy-bark and gum trees, black wood, wattles and not a single house within a mile' (McCorkell & Yule, 1999: 1).

During October 1854, Dr W.H. Harvey, an Irish botanist, visited the district, and recorded his impressions:

> On reaching the steep bank of the lake we looked down 200 feet into what must have been an extensive crater in old times, but now is partly a lake and partly a marsh. In the midst is a wooded island, rising like a cone 300 feet above the lake. There are two or three summits in which there are said to be small craters. The borders all around the lake have similar marks of volcanic origin and all are beautifully wooded. We had only time for a hurried scramble down the steep side of the lake; and the ground being covered with rich grass I got but few flowers, but among them was a beautiful Ajuga Australia, a fairy violet, a nettle and an indigo. On the waters of the lake myriads of a little floating fern, looking like duckweed, were swimming. The name is Azolla. For its size which is only inches across, it is extremely pretty (Lucas 1933: 187 in Downes, 1961: 31f).

Tim Bonyhady reported that 'Pteridimania'–as the fern craze was known from the 1850s in both England and Australia – was fuelled by the size of Australian ferns, which convinced colonists and visitors alike of the 'comparative insignificance' of their English counterparts. A rare dissentient was Irish botanist Dr W.H. Harvey, who failed to satisfy his sister's 'fern fever' while visiting Tower Hill in Victoria's Western District in 1855. But the fault was with Harvey. Had he made his way from the outer banks of the volcanic crater to the islands in its middle, as James Bonwick did two years later, he would have found 'gigantic ferns' forming 'delightful bowers and alcoves' (Bonyhady, 1987: 104).

In 1855, James Dawson, of Kangatong, commissioned German landscape painter Eugen von Guerard to paint Tower Hill (see Fig.8.1). Bonyhady has noted that 'Like many successful pastoralists, James Dawson wanted paintings of both his own property near Port Fairy and the most notable feature in his local landscape–Tower Hill, which he described as 'one of the most beautiful and interesting specimens of an extinct volcano in all Victoria' (Bonyhady, 1987: 76). The painting entitled 'Outlook' depicted a variety of flora and fauna which had virtually disappeared by the end of the 1860s due to the clearing practices of European settlers who used the local soils for agriculture and who grazed cattle on the islands.

> The people who commissioned von Guerard to paint homestead portraits were, of necessity, successful landholders, usually with substantial if not grand houses. The owners of small houses and bush huts typically could not afford to buy oil paintings. Nerveless such, houses, especially when set against a background of virgin forest or dramatic mountains were a popular subject

with many colonial artists because they could be used to convey the idea of the pioneer in the wilderness, the taming of the land (Bonyhady, 1987: 76).

Von Guerard's preferred subjects were fern tree gullies, waterfalls, rocky coastal scenes and views from mountains as well as sweeping views from mountain tops. In these pictures he was concerned to convey the grandeur of Australian scenery. He also sought to convey the magnificence of God's creation; as he put it: 'to catch now and then a glimpse of the divine poetical feelings' inspired by nature (Bonyhady, 1987: 357). But most of all von Guerard 'wished to paint as closely as he saw the details and effects of nature' as he stated:

> If ever [he] could succeed to paint Australian scenes to make them delightful illustrations for treatises of botanical or geological features of the colony then he would be convinced that for the future his paintings would be a greater value, where it would be doubtful if those which can be taken equally well for a misty English or an Australian landscape will have the same future (Bonyhady, 1987: 357).

His painting was later to feature prominently in the Tower Hill story and since 1961 has been used as the basis of a revegetation program entailing 300,000 plantings.

The educational inspector, James Bonwick, visited the district in 1857 and in his 1858 publication he gave a detailed account of the locality. He was clearly impressed with the wonders of Tower Hill:

> Tower Hill is the most interesting locality I ever visited, and I am indeed astonished that the people of Melbourne and Geelong generally should be ignorant of its wonders. It is an island or rather three islands together encompassed by a wall in the shape of a circular ash-bank, with water, a moat, inside and around. Something similar is known in the Phillipine islands, where an island of cinder and scoria springs nine hundred feet high from the body of Lake Taal. Tower Hill district is essentially a farming one. Immense crops are raised, though the land is chiefly leasehold portions of Rutledge's survey. Some portions of the Tower Hill bank have been sold lately. The settlers appear to be Irish or Scotch, there being few English. There is no township, though a road-side inn is erected for the convenience of travellers (Bonwick, 1858: 70).

> The surveyors had some time before constructed a crude bridge across the swamp between the banks and island, in order to transport a suitable pole for the summit of their Trigonometrically [sic] Station. The straw which they had spread over to fill up the gaping chasms has furnished many meals for straggling bullocks by the lake; so that in passing I had to make sundry leaps over and into moist places. On my first morning visit I ascended the Mount, being accompanied by the Rev. Mr Baker, an old traveller, and an expert mountaineer. Upon the next occasion, when alone, I walked around the base of the island, which with the indentations of its bays, I should imagine to be nearly five miles in length, although it is said to contain but eight hundred acres (Bonwick, 1858: 75).

> There was one perpendicular block of cinder, thirty feet in height, presenting some remarkable aspects of the volcanic agency; the sulphur, if any, had sublimated. Well might the natives exclaim as they saw this,–'Big one fire sit down along here.' What a gorgeous sight must have been witnessed by unsentimental kangaroos of old when Tower Hill was in full blaze! Doubtless

> the hopping tribe were peacefully nibbling when rumours of war's alarms arose. The rumbling of earthquakes shook the rock below. Steamy vapours pierced the rust of limestone (Bonwick, 1858: 76).

Although Bonwick spent little more than two hours exploring Tower Hill, his comments are significant for in his closing remarks he called for the reservation of Tower Hill:

> A stroll among the gigantic ferns of the valley, or a ramble among the cones and craters, has peculiar attractions. But these are not comparable to the winding path at the foot of the basaltic rises close to the lake. There the graceful Fern tree waves, ... almost tropical reeds rustle in the breeze, ... leafy shrubs and trees form delightful bowers and alcoves, and tender emotion in suitable company, will receive as genial and as rapid a development. The various impediments arising in the walk, such as the straggling roots, little inlets, tangled foliage, and jutting rocks, will only present occasions for the office of service, and the acceptance of kind assistance. Let the few who value sentiment in the colony [who] sympathize with nature, who love an undisturbed communion with the grand and sublime, join one and all in securing for themselves and posterity the authorised declaration, that Tower Hill shall be an everlasting reserve (Bonwick, 1858: 77).

The *Ballarat Star* published an account of a visit to the site in December 1869:

> The steep pathway down to the lake and the crossing to the island present an unusually lively aspect as gaily dressed groups were making their way to the island. A tent was erected in which the refreshments were stored under the direction of Mr Maltby of Koroit, who supplied them; but these were largely supplemented by contributions of hams, roast lambs, fowls and cakes from the ladies of members of the society. The officers of the Warrnambool St. Patrick's Society were specially invited on the occasion and were accompanied by a number of their leading townsmen. Among the visitors were some of the leading mercantile men of Belfast [Port Fairy]; and it is almost needless to say that numbers of the farmers of the district and their families joined in the festivities. About half past four, the officers in their regalia (green velvet collars richly trimmed with gold lace), accompanied Mr Deegan to the summit of the island, followed by a large assemblage (*The Ballarat Star*, 17/12/1869).

The Australian Sketcher with Pen and Pencil (1/5/1877) gave an account of Tower Hill for its readers:

> Tower-Hill, half way between Warrnambool and Belfast, is one of the most remarkable extinct volcanoes in the colony, and, according to Mr Rawlinson, who has studied the geological formation of that district, has played an important part in forming the coastline about Warrnambool. The following descriptive particulars are from *The Argus* report of a paper read by Mr Rawlinson before the Royal Society on the 14th June last: "Tower Hill is an extinct volcano of great extent and is so named from the likeness to a peak or tower of the central cones of eruption, which form an island in the middle of what is now a large lake of water. The circumference of the main crater is from five to six miles, and although not more than 320ft deep at present, it was when in activity and before the formation of the cones of eruption a yawning chasm, of probably 600ft to 1,000ft. ... Some years ago the cones were covered to the summits with trees, but over the greater part nothing but the trunks now remain. The high peaks are seen from a considerable distance at sea, and form a conspicuous landmark for vessels making the ports of Warrnambool and Belfast.

The Rev. H.M.R. Rupp lived at Koroit from 1876 until 1884, from the age of 4 to 11 years, and in his reminiscences he recalled how when a child, 'My father used to lead visitors up to the brow of the gently rising hill behind Koroit with their eyes closed. What exclamations there were when they were opened on the summit!' (Downes, 1961: 76).

The Illustrated Australian News reported on Saturday 29 June 1881 'The great object of interest to the traveller on the way to Belfast is Tower Hill., the highest peak in a cluster of hills that rise from Lake Koroit which is itself the bed of an extinct crater. Looked at from any point of view this wonderful formation is a striking and interesting example of nature's handiwork. What is now the lake was once the basin of the volcano. In many parts the lake is shallow and a wild rank vegetation springs out of the water, making it a glorious haven for lacustrine birds'.

The *Illustrated Australian News* (21/3/1883) confirmed that the Western District of Victoria has many claims to the picturesque, and noted that 'the well-known Tower-hill is one of the prettiest places that can be visited by the excursionist'.

> A drive from Warrnambool on a well-made and even road of about 9 miles, passing over the Merri-bridge, through the village of Dennington, leaving Werronggort, the estate of Mr Thomas Rutledge, on the left, and Farnham, that of Mr. William Rutledge, to the right, brings the visitor to that magnificent piece of landscape scenery known as Tower-hill Lake and Island, nearly half way between Belfast and Warrnambool. This island is estimated to contain 1200 acres. It rises to a height of 350 feet and is separated from the main land by the lake which varies in width from 300 to 1000 yards. The lake is circular in form and has evidently been formed by volcanic action; small volcanic craters may also be seen on the island. Excepting a small portion on the west side, the lake is surrounded by steep banks rising to the height of from 200 to 300 feet. The island has been permanently reserved, and a committee of management for its improvement has been appointed. The locality is a place of popular resort with those in the neighbourhood, and picnic parties are frequent during the summer season (Figure 8.2).

Julian Thomas aka 'The Vagabond' was another who visited the natural attraction and an account of his impressions was published in *The Argus* of Saturday 13 December 1884:

> I have heard a good deal of the drive around Tower Hill. I am not at all disappointed with my personal inspection of the locality; indeed I question if the natural beauties and material advantages of the country have been sufficiently magnified. ... A 13 miles drive from Warrnambool, and we are on the banks of Tower-hill lake. This is six miles in circumference, deep in some places, and abounding in eels; sedgy and marshy in others the banks bordered with reeds and dead gum trees. The extinct volcano is an island or rather three islands in the centre of this lake or moat which has been formed by the wall of ashes thrown up by the active crater, the accumulation making a circular bank almost as high as the volcano itself. This enormous fosse first became a marsh by the drainage from the surrounding eminences, and they many years back, Mr McCarthy having thrown the water back by an embankment at its outlet towards the Moyne, the lake as it now is was formed. The island is a public reserve of some 600 or 700 acres, a temporary crossing place making it available for tourist and picnic parties. Climbing the hills, one can study the old caters or inspect the small pool–a miniature edition of "Paradise Lake" on the island of Wanaka New Zealand. Or strolling through the little plains, valleys and dells one can speculate what a

> charming place this must have been before the larrikins from a distance amused themselves by setting fire to the grass and dry ferns. All the timber has been killed by this and the dead gum trees, gaunt and weird are, perchance in accord with the ancient spirit of destruction – the fire which was belched forth. But to the lover of nature this is a sad chance since when in 1857 Mr Bonwick described the island "there the graceful fern-tree waves; almost tropical reeds rustle in the breeze, and leafy shrubs form delightful bowers and alcoves". The geologist of the present day will however be fully satisfied with his visit to the island of Koroit and the opportunities there given for studying the varieties of volcanic action.

In 1886 English geologist, Dr John E. Taylor, visited Australia and published an account of his holiday in Australia. Taylor (1886: 193) wrote very positively of Tower Hill: 'It is trenched all round with water, which occupies the depression, and thus gives to this hill a most remarkable scenic appearance. Tower-hill alone is worth going to see'. Another favourable account was published in *The Argus* (12/8/1891):

> The beauties of the Tower Hill Lake an extraordinary volcanic formation on the sea coast between Warrnambool and Port Fairy –are vast. Many of the trees on the hill have been burned down or otherwise destroyed and the place has now a comparatively desolate appearance, even the wild fowl having deserted it. The borough council of Koroit is stirring in the interests of sport, and the secretary of Trade and Customs has received communication from Mr Richard Laffan, the local town clerk requesting that a proclamation should be issued prohibiting shooting on the lake. Mr Laffan states that the lake was formerly the resort of all species of aquatic wild fowl, but sportsmen have succeeded in driving them away.

8.2 Second Phase: Framing and Elevation 1866–1892

MacCannell (1976) stated that Elevation occurs when an object is put on display– in Tower Hill's case this is represented by its opening up for visitation. Framing is the placement of an official boundary around the object of interest. Two types of framing occur: protecting and enhancing.

The Bolden brothers, Armyne, George and Lemuel, were the first white settlers to use Tower Hill for grazing. They held licences for extensive runs in the area, including open forest and low flats of the formation. Tower Hill continued to be licensed for grazing runs until the 1850s, when Crown land allotments around the formation were sold. Many of these new freehold blocks extended to the rim of the crater and down the bank to the water's edge. Only small areas of Tower Hill remained in public hands in the period before the site was declared a reserve for public purposes in 1866.

In the early years of European settlement much of Tower Hill's natural vegetation was cleared and the land used for farming and quarrying. By the 1850s blocks of land were being leased and sold right down to the edge of the lake and land use became more intensive. The remaining timber was cut down and used for fuel. Early settlement at Koroit was scattered around Tower Hill, with two churches and a cemetery located on the south and south-eastern banks. Rutledge's Special Survey to the south of Tower Hill, where farming land was only available through leasehold, restricted development in that direction. Settlement to the north eventually became the township of Koroit. By the end of the 1850s more intensive use of the site was underway. The banks, including the steeper slopes, were cleared and planted with crops. A regime of clearing and burning was introduced, to promote pasture for livestock and areas for cultivation. Any remaining timber was removed, for fuel or firewood purposes. This use of the site was not restricted to the freehold land but included the island and those parts of Tower Hill which were reserved in 1866. The leased Crown land was subjected to frequent burning, to control the bracken and scrub regrowth, and rabbits flourished in the scrubby vegetation.

The Argus (16/9/1865) reported on the progress surrounding a request for a reservation at Tower Hill:

> The Minister for Lands acceded yesterday to a request made by a deputation from the residents of Koroit that a site should be reserved for a district recreation ground and botanical garden. Mr Grant also accepted the nine gentlemen who had been nominated as trustees. He likewise expressed his willingness to allow the island in Tower Hill Lake to be used for acclimatisation purposes, but he intimated that a formal written application would have to be made as it would be necessary to impose conditions that timber should not be cut, and that rabbits and other burrowing animals should not be placed on the island.

Although there had been interest for many years in preserving Tower Hill, it was not until 1866 that the island, the lake and the two-fifths of the banks which still remained as Crown Land were declared the Tower Hill Acclimatisation Reserve, under the control of the Tower Hill Acclimatisation Society. The public land reservation formed an area of 597 hectares. The Tower Hill Acclimatisation Society was expected to be self-sufficient and in order to achieve this they raised funds for fencing and the appointment of a Ranger. The society allowed timber extraction, grazing and clearing and let people live on part of the reserve, further contributing to the degradation of the vegetation. However, the members believed in its 'improvement' through introduction of European birds, animals and plants which later brought disastrous results to the reserve. In keeping with its aim to introduce exotic plants and animals they planted exotic trees and introduced rabbits, goats, jungle fowl and pheasants. The first permanent crossing to the island was completed in 1867. Crops of this period included mangolds (a kind of beet) planted on the island, potatoes and onions. For many years numbers of poor families had lived on the sloping banks, building small shanties or living in tents. Many of these had vegetable patches, raised cows and sometimes rented their blocks from the Acclimatisation Society.

In February 1868, a caretaker was engaged at a salary of £1 2/6 per week. He and his wife lived on the island in a tent for a year. His duties were to burn off the underwood, to care for stock, to keep persons from trespassing and preventing them from removing any of the animals and fowls that were placed on the reserve (Downes, 1961: 46). These included two angora and six common goats, five jungle fowl, three pheasants and two rabbits. It was commented at the time that the goats had effectively cropped all the young gum trees. For a time the island was used as a common where the fee for pasturing a horse was 1/- per week, a cow 6d, with foals and calves under six month free (Downes, 1961: 5). A survey of 1870 described 11 huts located on Crown land.

In November 1871 the Board of Land and Works approved the proclamation of Tower Hill Island of some 400 acres to be a reserve for the preservation and growth of timber called the Tower Hill Island State Forest, gazetted on 17 May 1872 (Downes, 1961: 54). On 8 September 1873, 1,475 acres comprising the Tower Hill lake and island were permanently reserved as a site for public recreation purposes (Government Gazette, 25/7/1873, p.1319, and 19/9/1873, p.1651).

Quarrying on Tower Hill began in 1878 with the removal of a small amount of tuff for what seemed a good reason–to make paths within Tower Hill Cemetery. By 1894 there were seven quarries around the site, including three within the Reserve. Quarrying on the island finally ceased in 1941, but continued on the banks. Over the next eighty years several hundred thousand cubic yards were removed until quarrying was stopped within the rim of Tower Hill in 1961. One private quarry on the southern rim was still being worked in 1966, although efforts were made to prevent the destruction of this unique area. Exposure of the volcanic layers brought about through quarrying has enabled scientists to study the Tower Hill formation more easily.

In the 1880s Koroit Council, then Committee of Management, permitted local ratepayers to depasture livestock on the Reserve. Cultivation was still permitted, on condition that the cleaning of scrub and bracken continued. There were still at least six families living on the lands in 1883, growing potatoes and oats, maintaining vegetable gardens and grazing livestock.

The Argus (11/7/1882) reported ‘a telegram made from our Koroit correspondent, published in *The Argus* of yesterday stated that the Lands department had handed over the control of the Tower Hill reserve to the Belfast Borough Council. There appears to be a wrong impression in the district as to the action of the department. The management of the Tower Hill reserve has not been interfered with. The mistake appears to have arisen from the fact that the Belfast town common has been placed under the control of the local council’.

In its original state Tower Hill abounded with vegetation and wildlife. However it is obvious from just a few comments recorded by early pioneers and visitors, how quickly European settlement transformed the area. Tower Hill had become a monument to the general destruction of wildlife habitat caused by intensive and unplanned agricultural development of much of Victoria. Historically and pictorially, these changes and their sequences are documented very well at Tower Hill. The influx of settlers about this time, many from the goldfields, led to a rapid change in the appearance of Tower Hill. Paintings and photographs from the earliest period in 1887 show the method of destruction; burning, grazing and cutting for firewood had by this time removed all timber of any size. After good wildlife conditions were replaced by a uniformly poor vegetation of bracken and grasses, there was a dramatic reduction in the numbers and variety of wildlife. (Downes, 1961: 5)

When James Dawson revisited Tower Hill in 1891 he was disgusted with what he saw, a landscape degraded by clearing, grazing, timbering, and the introduction of exotic plants and animals. He discovered that it no longer resembled the landscape painting by Eugen von Guerard, which he had commissioned in 1855, and he publicly protested. Tower Hill, he maintained, was 'intended by Nature to be a public park, and a place of pleasure for the people of the district' (Bonyhady, 2000: 341). Dawson published his impressions in the *Camperdown Chronicle* on Tuesday 24 February 1891:

> Sir.—In the early days of this colony there was to be seen between Port Fairy and Warrnambool, one of the most beautiful and interesting specimens of an extinct Volcano in all Victoria, "Tower Hill and Lake," then in their primitive state, attracted my attention so much, that, fortunately for future generations, I commissioned a celebrated artist to paint the scene in oil on a large scale, and he carried out my wishes faithfully and beautifully. On visiting the cone lately, I was amazed and disgusted to find everything altered; the fine trees on the cones and in the craters of the island all gone excepting half a dozen or so. But what was a thousand times worse than this ruthless destruction of ornamental timber, the larger portion of the lake is made into a seething, stinking, mud-pool, full of fermenting malarious matter, sufficient to poison the whole neighbourhood, and which to prevent any sensible person settling near it. Our inquiry was informed that this disgraceful and unsavoury pollution of a pure lake was the work of the inhabitants of the village of Koroit, who, to save trouble in getting rid of their foul matter by its legal and natural, excuse of drainage, deliberately shunted by their neighbours by means of a tunnel through the bank of the lake; thereby polluting its water, and causing a state of matters which very soon will prove fatal to the health of the locality. Since I visited the "mud pond" I have been informed that the people of Koroit derive their chief supply of water, by means of a pipe from a pump well sunk in the inner bint of the lake, close to the mouth of the tunnel. If this information is correct, who can be astonished at the spread of malarious diseases?—Yours, etc, JAMES DAWSON.

He published a second letter in the *Camperdown Chronicle* on Tuesday 10 March 1891:

> Sir, While lately in Melbourne I received copies of the Camperdown Chronicle of the 24th Inst., entertaining my letter on the disgraceful treatment of "Tower Hill and Lake." in the Western district, by the ruthless destruction of the timber on the island, and the discharge of the fermenting filth and sewage of Koroit by tunnel into the lake, thereby producing such a state of matters us will shortly produce a pestilence in the neighbourhood, besides damaging the water supply of Koroit, which I was informed on good authority is obtained from a well within a few yards of the inner mouth of the tunnel. Determined that this important matter should be brought under the notice of the proper authorities and investigated, I had interviews with the Secretary of Lands, and Secretary of the Board of Health, and on stating my own observation on the state of the lake, accompanied with copies of my letter published in the Chronicle, these gentlemen gave me an assurance that there would be an inquiry, which I hope will be hold while the hot season lasts. Yours, etc. JAMES DAWSON.

Dawson published a third letter in the *Camperdown Chronicle* on Tuesday 7 April 1891:

> SIR,—On the 24th ult., you kindly published a letter from me with the above title, setting forth unwarrantable destruction of the fine trees on the island, but what was a thousand times worse, the diverting the drainage of Koroit into Tower Hill lake, by means of tunnel cut through the bank, thereby converting it into a stinking, seething mud-pool. This was replied by the Koroit Sentinel in leading article as follows: "We fail to see what reason there is for making such an unfounded attack on the inhabitants of Koroit, as has been done in the letter alluded to, and we advise persons who wish to attack the district, to take the trouble to get some information about it, and not be in haste to print statements that are as absurd as they are untrue. In accordance with the advice of the writer of that article I again very recently visited the locality, and I found abundant corroboration of my information concerning the drainage of Koroit. It appears to be unquestionable that the only drainage of miles of roads, streets, and by ways, is directed into Tower Hill Like, and nowhere else, with the result, that the lake is not only a seething mass of fermentation at the end next to the tunnel, but the water at the other end is very bad, where pure water, might have been expected. The writer then says that "it is only after heavy rain that any water runs through the tunnel, and then only clean water," and as a proof of its purity, he mentions that "a sample of the water as it ran through the tunnel into the lake was forwarded to Melbourne in 1886 for analysis by the inspector of nuisances, and done so privately, that neither the borough councillors nor any of the officials, except the inspector of nuisance had any idea that the analysis was being made." The report of the analyst stated that "the water was pure rain water, with a certain amount of mud in solution, but no offensive matter." On this as above described "so privately" obtained sample of water in 1886 said 'report' was founded, and is now quoted to prove the purity of the drainage in 1891, and notwithstanding the increase of population and traffic since 1886 is still made an excuse for the perpetuation of a nuisance. For further information on the treatment of Tower Hill and Lake, I refer the writer of the leader in the Koroit Sentinel to previous correspondence on the subject.—Yours. &c, JAMES DAWSON, Local Guardian of Aborigines, Camperdown.

Dawson was eighty-four and in poor health in 1891, and must have expected that his letter would be read in Koroit, where the local *Sentinel* newspaper admitted it knew nothing of von Guerard's painting, but otherwise accused Dawson of making 'not one true statement'. The *Sentinel* maintained that the trees on the islands had been destroyed by fire rather than felled. There also was 'not the slightest miasma from the lake': the drain simply carried storm water, and the local council exercised the 'greatest care' to ensure 'no offensive matter' ran into it (Bonyhady, 1987: 343). By then, Dawson had gone directly to the colonial government. On a visit to Melbourne, he arranged interviews with the Secretary of Lands and the Secretary of the Board of Health, and persuaded them to investigate the lake (Bonyhady, 1987: 344). This activism came to nothing. Local health officers replied that the state of the lake was due to a 'natural drying-up process'; the drain carried only storm water after heavy rain and no one living on the lake bank had suffered a serious illness in seven years. But within a few months, local residents were also protesting about 'offensive matter' going into the drain. Later accounts make clear that the town's waste sometimes reached the lake, justifying Dawson's insistence that it was a 'seething swampy lagoon, supplied through two tunnels with the sewage of the town of Koroit, and the numerous farms in the district' (Bonyhady, 1987: 344).

When ownership of the *Sentinel* changed in late 1891, its new proprietors made much of the 'beautiful natural monument' at Tower Hill because it was one of Koroit's 'most popular places of resort', and they fancied that it could assist Koroit to be chosen as the new national capital when Australia became federated. They also acknowledged that Tower Hill's 'beauty and loveliness' had, at least temporarily, been 'ruined': its wildfowl driven away by professional shooting on the lake needed to be restricted, if not banned, and urged the local council to restore Tower Hill's trees (Bonyhady, 1987: 345).

Six month later, the local member, Sir Bryan O'Loghlen, introduced a Bill to vest the lake and islands of Tower Hill in the council, which he hoped would 'restore the spot to its pristine beauty'. O'Loghlen underlined this idea by calling his legislation the 'Tower Hill National Park Bill'. The press in Koroit and Warrnambool agreed that national parks were the only means of persevering 'the indigenous trees, shrubs and flower of our island continent'. Yet the vesting of Tower Hill in the council was never likely to lead to its protection, let alone restoration, because Koroit was accustomed to exploiting it. Although formally a reserve for public recreation, in 1891 it served as a town common where every ratepayer could depasture five head of stock. When the *Sentinel* called for the replanting of Tower Hill, the council responded by setting fire to one of the islands. Its object was to remove the undergrowth, which reduced the feed available for grazing and was 'a great harbour for rabbits'. But the blaze left 'hundreds of burning stumps' (Bonyhady, 1987: 345).

In 1892, it became Victoria's first National Park in recognition of its unique features, but wasn't managed as a National Park as people still continued to farm and quarry there. The Tower Hill National Park Bill was passed on 5 December 1892, vesting Crown land at Tower Hill in the control of the Borough of Koroit. The area was to be maintained as a park for public recreation purposes. This was the first 'National Park' in Victoria, but it was not in the style of present day national parks. O'Loghlen's Bill was also flawed. Rather than defining a 'National Park' as a place to be preserved or rehabilitated, it simply vested Tower Hill in the council 'as a park for public recreation purposes'. The bill also did not address how the restoration of Tower Hill might be paid for; the council's receipts from grazing on the island were insufficient to pay for a ranger, let alone clear the islands of its rabbits, which had already survived fires, a Saturday shoot by local residents and burrows filled with poisoned gas (Bonyhady, 1987: 345).

Although the Tower Hill National Park Act 1892 stipulated that Tower Hill could be used for 'no other purpose whatsoever' than public recreation, the Council turned it into a quarry, using its scoria as 'road metal' and its tuff for gravel. Although there were many other deposits of scoria and tuff nearby, Tower Hill became the main local source for road-making. By 1894, there were seven pits–three in the national park, the other four on private land–leaving great gouges in the crater and islands. Koroit was probably also to blame for the disappearance of the lake on the eastern side of the islands. The water level in the lake started dropping almost as soon as the council began tapping the main spring in 1890; then the lake dried up for extended periods before finally disappearing altogether during the drought of 1914, allowing stock to graze unimpeded on the islands (Bonyhady, 1987: 345). Tower Hill remained firmly under the control of the Koroit Council, which continued to allow grazing and quarrying in the park for many years.

8.3 Third Phase: Enshrinement 1892–Present

MacCannell identified enshrinement as the third phase. This occurs when there is an increase in tourism numbers and the reputation of the attraction is enhanced 'when the framing material' itself becomes integrated into the sight.

The *Horsham Times* (17/12/1895) reported on the state of tourism at Tower Hill in the 1890s:

> TOWER HILL: A drive to Tower Hill is usually the first proposition to a tourist, and it is one which will repay the journey. The visitor leaves Warrnambool by the Belfast Road, passing through the township of Dennington, about three miles distant. After Dennington bridge has been crossed, the magnificent pastures of Farnham Survey are passed, occupying the road on either side. Passing the hilly and sinuous road through the estate, the visitor detects his emergence into the volcanic region of Tower Hill. In the cuttings of the road may be seen layers of ashes belched forth by the volcano in its igneous activity. The Vagabond says of this piece of natural scenery: A 10-mile drive from Warrnambool, and we are on the banks of Tower-hill lake. This is six miles in circumference, deep in some places, and abounding in eels; sedgy and marshy in other, the banks bordered with reeds and dead gum trees. The extinct volcano is an island, or rather three islands in the centre of this lake, or moat, which has been formed by the wall of ashes thrown up by the active crater, the accumulation making a circular bank almost as high as the volcano itself. The island is a public reserve of some 600 or 700 acres, a temporary crossing place making it available for tourists and picnic parties. Climbing the hills, one can study the old craters or inspect the small pool, a miniature edition of "Paradise Lake," on the island in Wanaka, New Zealand. From the highest point of the island the site of the old trignometrical station, the most extensive view is obtained, but from the road round the lake bank one has just as grand a *coup d'oeil.*[33] The waters of the lake are 25 feet below us: the banks all covered with verdure, sheep grazing to the very edge. From the Grampians to the white-fringed seashore, on all sides distant outlines of hill and mountain, enclosing it glorious view of plain and valley, of flowing streams, wooded dells, green meads, fertile paddocks, with pleasant homesteads and prosperous townships dot- ted here and there. A grand magnificent view this I question if I shall see such another panorama in Australia. From Beechworth, from the Alps, from Mount Major and Mount Hope, I have been privileged to view something of Victoria, but the country around Tower-hill possesses, besides the adjuncts of mountain and plain, wood, water, and sea, a special interest of its own."

---

33 A French expression that means a glance, glimpse, or a quick look.

> From Tower Hill, the visitor should drive on to Koroit, a flourishing town in the centre of the rich agricultural land, and a pleasant half-hour can be spent by visiting the Koroit Botanical Gardens.

The degradation of Tower Hill was exacerbated by the first significant attempts at tree-planting, which added Monterey pines to the crater's bank. After the town's United Friendly Societies' Committee persuaded the council to obtain trees from the State Forests Commission, local school boys planted 600 pines in 1924, which flourished behind a protective fence. The schoolboys knew that 'when the early pioneers first came to settle near Tower Hill' it had been 'a real Australian wonderland', its banks covered with 'beautiful trees-ferns, hazel and musk bushes'. Despite their very different plantings, they saw themselves as recreating this 'well-wooded' past and enhancing the future. In an essay in the local school magazine one of the boys declared, 'We shall watch their growth with interest....for we feel sure that what we are doing will be of great benefit to those who come after us'. Other local residents wanted native trees. P.E. O'Grady, a Warrnambool accountant, exclaimed in 1937, 'What a wonderful scene could be recreated if Tower Hill...was replanted with native timber!' He thought the Warrnambool Rotary Club would help because it was 'interested in all tree-planting schemes' (Bonyhady, 1987: 355). However years later the Australian Natives' Association planted a stand of manna gums, the staple source of food for koalas in southern Australia. The Association wanted to establish a chain of koala colonies through Victoria, where the fur trade, habitat destruction and disease had reduced the koala population to perhaps 500 in 1925. But rabbits burrowed under the protective fence and killed almost all the trees within a year, so the council reverted to planting pines and willows.

When the National Parks Act was introduced in 1956, Tower Hill was omitted because of its degraded condition. It had been virtually stripped bare and little wildlife remained. The Warrnambool Field Naturalists Club held its inaugural meeting in March 1958. In February 1959 the Field Naturalists requested one acre to be planted in native trees which was granted. It was the initial step in a program to restore Tower Hill to its former beauty (John, Moore & Moore, 1992).

In May 1960, Victoria's new State Wildlife Reserves Investigation Committee approved to transfer Tower Hill to the Fisheries and Wildlife Branch as Victoria's seventh Game Reserve. Four months later this decision was law and Victoria's first national park no longer. Alienation of parts of Tower Hill began at an early date and since 1961 it has been necessary for the Government to purchase back more than 110 acres of the banks. A considerable length of the bank of Tower Hill remains in private control (John et al., 1992: 8).

The Fisheries and Wildlife immediately stopped quarrying and grazing within the reserve and appointed two bailiffs, who began poisoning foxes, eradicating noxious weeds such as boxthorn and thistles, and attacking the islands' rabbits, which were the primary obstacle to the growth of new trees. They were also quick to try to control the use of private land both within and outside the crater. In 1961 it infuriated local councils and landowners when the Fisheries and Wildlife persuaded the government to make an interim development order giving Victoria's new Town and Ocean Planning Board control over the area between Tower Hill and the ocean. While this order did not affect the quarries already operating on private land, it required the Board's consent for all new land uses, including subdivision and the opening of further quarries. At the same time, the Branch sought to buy up and close the existing quarries. By early 1962 it had acquired seventy-four aces on the craters eastern banks; by 1966 only one quarry was still operating (Bonyhady, 1987: 355).

The Department of Fisheries and Wildlife was quick to publicise its work. Even in 1960, before it actually took charge of the old national park, it published the first of a series of pamphlets by Downes in which he cast Tower Hill as 'a monument to...the destruction of vegetation and wildlife which resulted from settlement of Victoria by the white man' and emphasised 'the opportunity...to bring this area back to something approaching its original condition'. In 1962 the Department worked with the Country Roads Board and the Tourist Development Authority to construct a new lookout on the Princes Highway overlooking Tower Hill. Donations were received from 1962 to commission architect Robin Boyd to sketch plans for a natural history centre. The purpose of the centre was to show the public how wildlife reserves conserved wildlife and to provide details of the conservation work and the historical records of Tower Hill, and the early settlers who tried to preserve it. In 1964 the Department re-established access to the islands by rebuilding the causeways, which had been submerged in the 1950s (Bonyhady, 1987: 356).

Once Tower Hill was declared a State Game Reserve under the Wildlife Act in 1961, the Fisheries and Wildlife Division was responsible for the management and replanting programme which commenced in 1961. When Tower Hill was declared a State Game Reserve in 1961, the then fisheries and Wildlife Division of the Victorian Ministry of Conservation embarked on a restoration program based on Von Guerard's painting. The painting was so detailed that a naturalist, J Ros Garnet, was able to identify at least twelve plant species. The artist's vision was used as a template for the restoration. In the eighteenth century, landscape gardeners such as Capability Brown created artificial wildernesses based on paintings of Claude and Poussin. In Australia, scientists scrutinised von Guerard's brushstrokes and turned art into nature (Griffiths, 1991:21).

Max Downes serendipitously discovered the von Guerard painting in 1960 when he went to Melbourne to visit Mrs K.P.M. Winter, the daughter of Isabella Dawson, and granddaughter of James Dawson. The painting had been commissioned by James Dawson to show the vegetation 'as the Aborigines knew it and just before the Europeans ruin it' (Ryllis Clark, 1996). The fact that von Guérard was a botanist meant that he had a very good eye for the detail of the vegetation, and this has greatly aided the planting programme. Botanists studied the painting to identify plants, and replanting was based on species known or believed to have been indigenous to Tower Hill. For six years Von Guerard's painting was displayed at the Natural History Centre at Tower Hill. The Branch decided that the centre was too insecure to house the painting which jumped in value in the 1970s as von Guerard came to be regarded as one of Australia's finest colonial artists. As a result, the Branch decided to loan it to the Regional Gallery in Warrnambool where it was soon surrounded by new paintings of Tower Hill (Bonyhady, 1987: 363).

The Fisheries and Wildlife Department developed a planting scheme using von Guerard's detailed painting as a reference as it shows grasses and ferns on the island, and tea trees, wattles, she oaks, banksia and eucalypts on the cones with reeds and tussocks in the marshes.

By 1981, around 25,000 trees and shrubs had been planted with the help of school children and volunteers along with herbs, grasses and rushes. As the plants became established, introduced plants and weeds were removed. More recently, Parks Victoria conducted a pollen analysis at Tower Hill which helped identify other plants growing before European settlement (Parks Victoria, 2012).

Community involvement has been a critical part of the re-vegetation and redevelopment of Tower Hill over the last 50 years. It is an integral part of the local Warrnambool, Koroit and Killarney communities and local stakeholders feel strongly about its long term future.In 1992, 'The Friends of Tower Hill group' was established. The group meets for a working bee each month and is joined by Parks Victoria representatives. In 2012, *The Moyne Gazette* (2012) reported that 'The Friends of Tower Hill' is still going strong two decades later with an active membership of 10 as well as a number of other members who live outside of the area'.

In 2002, the Worn Gundidj Aboriginal Cooperative entered into a Commercial Services Agreement with Parks Victoria to provide visitation experiences, tourist information and saleable items at Tower Hill Wildlife Reserve. Worn Gundidj volunteers with their technical expertise and horticulture resources liaise with schools that are actively engaged in Tower Hill revegetation programs. Worn Gundidj also facilitated a Natural Heritage Trust program that resulted in some ten thousand units being planted and guarded at Tower Hill. On the tourism front they are product partners of the Great Southern Touring Route offering fun-filled and informative activities ranging from Kids Games to Aboriginal history and Environmental Science. Worn Gundidj Aboriginal Cooperative, with Parks Victoria, has revitalised Tower Hill. 'The most important thing at Worn Gundidj is to reinvigorate the culture and storylines of our people,' says John Collier, Executive Chairman of the Cooperative. 'We take a holistic approach to the area. Not only do we do interpretative tours, but we're involved in conservation, tree planting and looking after the natural habitat. Indirectly that transfers skills to our guides' (*Australian Traveller*, 4/3/2008).

8.4 Fourth Phase: Duplication

MacCannell's (1976) fourth phase in the development of attractions is that of 'duplication', when copies of the nucleus of the attraction, in this case the caldera, are made available through media such as paintings, photographs, and postcards.

The most famous duplication of Koroit or Tower Hill is the 1855 oil painting by Eugen von Guerard, entitled 'Outlook' which today is housed at the Warrnambool Regional Art Gallery. Printed reproductions of 'Outlook' were made by Thomas McLean of London and carried the caption 'View of Moroit [sic] or Tower Hill. Extinct volcano between Lady Bay and Port Fairy in Australia Felix. The original in the possession of James Dawson Esq. of Kangatong' (Downes, 1961: 32). Other significant duplications include a sketch by Bonwick in 1857 (Bonwick, 1858); a series of four paintings in 1867 by Daniel Clarke of Warrnambool (Downes, 1961: 37). For a more complete list of paintings and early photographs see Downes (1961).

In the case of Tower Hill, the interpretive centre, now known as the Worn Gundidj Visitor Centre, on the main island, duplicates the caldera in its design. The centre was completed around 1970 and was designed by Melbourne architect Robin Boyd. The building is designed in the shape of a volcanic cone and has displays of geology, flora, fauna and human history (including Aboriginal dress) of Tower Hill (see Fig.8.3). The Centre also sells souvenirs and products such as bags, napkins, tea-towels made by the local Worn Gundidj Aboriginal Cooperative. These products are also available online. The Centre is run by the Worn Gundidj Cooperative in conjunction with Parks Victoria.

8.5 Fifth Phase: Social Reproduction

For MacCannell, the final stage of sight sacralization is social reproduction which 'occurs when groups, cities, and regions begin to name themselves after famous attractions'. In the case of Tower Hill, social reproduction has occurred in the naming of a local accommodation business – 'Tower Hill House', Tower Hill Boutique Lavender and Oil Farm. The local caravan park employs both names: 'Koroit-Tower Hill Caravan Park'. In terms of the traditional Aboriginal name for Tower Hill, that of Koroit, this name has been reproduced in present usage in the nearby township of Koroit which dates from 1857 and the borough of Koroit; Koroit Art Gallery; Mickey Bourkes Koroit Hotel, and other businesses and organizations.

8.6 Tourism at Tower Hill

From a state-wide tourism view point, Tower Hill is generally regarded as a secondary attraction in the overall Great Ocean Road tourism product, particularly in relation to Port Campbell National Park. However, at a regional and local level, Tower Hill is a very important attraction. In terms of on-site self-guided walks, the reserve has five walks – each with different themes: Peak Climb – Orientation and geology; Lava Tongue – Boardwalk wetlands and wildlife; Hat Island – Habitat track revegetation; Whurrong Walk – Wetlands and habitat; and Journey to the last volcano – geology and wildlife. The reserve also offers picnic facilities, promotes wildlife viewing; and has the Worn Gundidj Visitor Centre that promotes Aboriginal and natural heritage of the reserve. Visitors are able to gain access to the park at all times, with the Visitor Centre being open at set times. The visitor centre is a not for profit organisation, the Visitors Centre is managed to ensure that visitors have a high quality experience and ensure that the facilities are well maintained. It is also possible to have guided experiences that centre on Aboriginal history; Wildlife encounters; Volcano and nature walks; Night adventure treks; and the centre is able to be booked to host birthday parties for children; and offers interpretive classes on understanding nature for children/school groups.

The Department of Sustainability and *Environment* (DSE) has the role of ensuring the sustainable management of natural and cultural resources for both current and future generations. Parks Victoria is responsible for the planning and management of Victoria's parks and reserves systems including Tower Hill. Both agencies realise that tourism can provide a vehicle for maintaining the natural and cultural assets, while meeting the needs of Victorian residents. It was suggested by Parks Victoria that the name Tower Hill State Game Reserve should be changed to either Tower Hill Volcanic Reserve or Tower Hill Heritage Park, as it was thought that a more appropriate name would be better for marketing, however, in 2013, the name is still Tower Hill State Game Reserve.

The Worn Gundidj Aboriginal Co-operative Ltd commenced operation in 1992 and along with Parks Victoria began the management of the Visitor Centre in the same year. Worn Gundidj Aboriginal Cooperative aims to achieve the following goals: Tourism–the operation of tours and a Visitor Centre at Tower Hill State Game Reserve near Warrnambool, Victoria; and Horticulture – a nursery that generates sales of potted colour and native plants, and environmental rehabilitation and services.

Tower Hill is also part of the UNESCO endorsed Kanawinka Geopark. The Kanawinka Geopark, named for *Kanawinka*, "Land of Tomorrow", is situated along a structurally controlling geological fault of the same name that extends from the Narracoorte Caves in South Australia into Western Victoria, before disappearing offshore at Portland.

In 1953, *The Argus* (5/5/1953) reported on the tourism possibilities at Tower Hill, and noted that local residents compared it with American canyons and New Zealand hot springs.

Table 8.2 presents a chronology of major developments at the site:

Tab. 8.2: An Historical Timeline of Tower Hill

Date	Event
30,000 BC	Volcanic eruption / formation. This was eventually followed by natural revegetation.
Until approximately 1840	Used by the Koroitgundidj people and their descendants
1802	First confirmed European sighting (Captain Baudin aboard the Geographe)
1840	First Occupation by pioneer graziers (clearing commences)
1855	Eugen von Guerard painted 'Outlook'
1892	Declared Victoria's first 'National Park' by local council (but not managed as such)
Mid 1950s	Grazing, quarrying, crop growing and rubbish dumping continue until the hills and island are bare and little wildlife remains. Duck shooting was also allowed. Extensive lobbying for revegetation including efforts by the Warrnambool Field and Game Club and the Warrnambool Field Naturalists Club.
1961	Declared a State Game Reserve under Wildlife Act. Fisheries and Wildlife Division responsible management. Replanting program commences. Over 300,000 trees planted with the help of conservation, community groups and schools. Tower Hill becomes recreation and conservation asset for the local community and for visitors. Site is significant for interpretation and education.
1969/70	Robin Boyd-designed Centre constructed. Series of interpretive displays subsequently installed.
1997	Parks Victoria became responsible for management. Managed under Wildlife Act.
2002	Commissioning of the Tower Hill Future Directions Strategy.

In December 2013, Worn Gundidj Aboriginal Cooperative announced the release of a new tourism package centred on Tower Hill, called an 'escape package', which is focussed on the region's Indigenous culture, landscapes, historic shipwrecks, volcanoes and native animals. The package includes a guided bushwalking 'walk with wildlife' tour through the reserve at night, which gives visitors the chance to experience Tower Hill's silhouetted crater and landscape at sunset as well as the night time nature walk (Tourism Australia, 2013).

8.7 Conclusion

Tower Hill as a tourism attraction has significant natural values for local and international visitors. It became a tourist attraction more for its natural significance rather than its Indigenous values, although visitors to the site are able to learn about the Indigenous people and their way of life. Tower Hill is one of the first natural features in Victoria to be given a European name 'Peak of Reconnaissance' in March 1802 by the French explorers Baudin and Peron. A nested caldera, it is one of the most recently active volcanoes in western Victoria, and a well as these significant natural values it has significant Aboriginal heritage values as represented in the detailed Aboriginal microtoponymy that has been recorded in the ethnographic record. The name Tower Hill is descriptive and is believed to date from 1838. Although European settlement and land use began from this time, portions of the caldera remained as Crown Land and in 1866 local residents were successful in their endeavours to reserve Tower Hill when the Victorian Government declared it the Tower Hill Acclimatisation Reserve. When schools inspector James Bonwick visited the site in 1857 he considered it sublime and grand and urged its reservation. Two years previously, local settler James Dawson had commissioned German landscape painter Eugen von Guerard to paint Tower Hill. Years later, in 1961, this most famous example of duplication of the site featured prominently in attempts to restore the vegetation of the island to that depicted in the von Guerard painting. In 1892, Tower Hill became Victoria's first National Park and was maintained as a park for public recreational purposes by the local shire council. However, continued environmental degradation of Tower Hill led to its omission from the 1956 National Parks Act. Local Field Naturalists were responsible for efforts to arrest the degradation and take active steps towards its environmental rejuvenation. In 1960 Tower Hill became victoria's seventh Game Reserve, and the government department responsible, the Department of Fisheries and Wildlife set about establishing a natural history centre and embarked on a restoration program based on von Guerard's painting (see Fig.8.4). This chapter has given an historical analysis of Tower Hill State Game Reserve, documenting its evolution from when Europeans inhabited the Reserve to the present day where it has been rejuvenated and become a favourite attraction for locals and visitors to the area.

References

Australian Plans Online. (1996). Retrieved January 2, 2013, from The Tower Hill Experience: http://anpsa.org.au/APOL2/jun96-2.html

Australian Traveller. 4/3/2008

Australian Traveller. (2012). Retrieved December 5, 2012, from Australian Traveller: http://www.australiantraveller.com/warrnambool/worn-gundidj/

Bonyhady, T. (1987). *The Colonial Image*. Chippendale: Ellsyd Press.

Bonyhady, T. (2000). *The Colonial Earth*. Melbourne: Melbourne University Press.

Bonwick, J. (1858). *Western Victoria; its geography, geology and social condition. The narrative of an educational tour of 1857*. Geelong: Thomas Brown.

Brady, A. (2012, October 18). Retrieved January 9, 2013, from Moyne Gazette: http://www.moynegazette.com.au/story/402890/friends-of-tower-hill-to-improve-facilities/

Camperdown Chronicle, 24/2/1891; 10/3/1891; 7/4/1891

Community South West. (2010, June 9). Retrieved December 5, 2012, from Worn Gundidj Cooperative: http://www.communitysouthwest.com/membership/worn-gundidj

Clark, I.D. (1990). *Aboriginal languages and clans: an historical atlas of western and central Victoria, 1800-1900*. Melbourne: Department of Geography and Environmental Science, Monash University.

Clark, I.D. (ed.) (2000a). *The Journals of George Augustus Robinson, Chief Protector, Port Phillip Aboriginal Protectorate, Volume One: 1 January 1839-30 September 1840*. Clarendon: Heritage Matters.

Clark, I.D. (ed.) (2000b). *The Journals of George Augustus Robinson, Chief Protector, Port Phillip Aboriginal Protectorate, Volume Three: 1 October 1841-31 December 1843*. Clarendon: Heritage Matters.

Clark, I. D., & Heydon, T. (2002). *Dictionary of Aboriginal Placenames of Southwest Victoria*. Melbourne: Victorian Aboriginal Corporation for Languages.

Clark, I.D. (2009). Reconstruction of Aboriginal micro-toponymy in western and central Victoria – case studies from Tower Hill, the Hopkins River, and Lake Boga. In H. Koch & Hercus, L. (eds), *Aboriginal Placename: Naming and re-naming the Australian Landscape* (pp. 207-221). Canberra: Aboriginal History Inc. & ANU E Press.

CPG. (2011). *Koroit Strategy Plan: Context Report*. Moyne Shire Council. Retrieved January 6, 2013, from http://www.moyne.vic.gov.au/Files/Draft_Koroit_Context_Report_March_2011_version.pdf

Dawson, I.P. (1870). The language of the Aborigines, letter to the editor. *The Australasian*. 19/3/1870.

Dawson, J. (1881). *The Australian Aborigines: the languages and customs of several tribes in the Western District of Victoria*, Australia. Melbourne: George Robertson.

Destination Recovery Services. (2007). Retrieved January 9, 2013, from http://www.destination-recovery.com/destinationlifecycle.html

Downes, M.C. (1961). *The History of Tower Hill to 1892*. Koroit: Fisheries & Wildlife Division, Tower Hill State Game Reserve.

Fisheries and Wildlife Department. (1983). *Tower Hill State Game Reserve*. East Melbourne: Ministry for Conservation.

Griffiths, T. (1991). History and Natural History: Conservation Movements in Conflict. *History and Natural History*, 24(96): 16-32.

Horsham Times, 17/12/1815

Illustrated Australian News, 21/3/1883

Learmonth, N.F. (1934). *The Portland Bay Settlement being the history of Portland, Victoria from 1800 – 1851*. Portland: The Historical Committee of Portland.

Kanawinka Global Geopark (2010-2012). Retrieved January 9, 2013, from: http://www.kanawinka-geopark.org.au

MacCannell, D. (1999). *The Tourist*. Los Angeles: University of California Press.

McCorkell, H. A., & Yule, P. (1999). *A green and Pleasant Land–A history of Koroit*. Warrnambool: Collett, Bain & Gaspar.

John, C., Moore, H., & Moore, R. (1992). *Redressing the Volcano: the Restoration of Tower Hill*. Tower Hill: Tower Hill Press.

Mulder, J.F. (1909). Stone Implements of the Natives of Victoria, and their Age. *The Geelong Naturalist*. Second Series. 4(1): 8-14.

Museum Victoria. (n.d.). Retrieved January 9, 2013, from Collections: http://museumvictoria.com.au/collections/themes/1948/eugene-von-guerard-artist-1811-1901

National Gallery of Victoria (n.d.). Retrieved January 9, 2013, from: http://www.ngv.vic.gov.au/whats-on/exhibitions/exhibitions/eugene-von-guerard

On My Doorstep. (2010). Retrieved January 2, 2013, from http://www.onmydoorstep.com.au/heritage-listing/14278/tower-hill-state-game-reserve

Parks Victoria. (2012). Retrieved December 5, 2012, from Tower Hill Reserve: http://parkweb.vic.gov.au/__data/assets/pdf_file/0005/315662/Park-note-Tower-Hill-Reserve.pdf

Parks Victoria. (n.d.). Retrieved December 5, 2012, from Tower Hill Reserve–History and Heritage: http://parkweb.vic.gov.au/__data/assets/pdf_file/0007/315547/Heritage-story-Tower-Hill-Reserve-history-and-heritage.pdf

Parks Victoria *Tower Hill-Culture and Heritage*. (2012). Retrieved December 5, 2012, from: http://parkweb.vic.gov.au/explore/parks/tower-hill-w.r/culture-and-heritage

Parks Victoria. (2012). *Parks Victoria*. Retrieved December 5, 2012, from ParkWeb: http://parkweb.vic.gov.au/explore/parks/tower-hill-w.r/things-to-do/take-a-self-guided-walk-at-tower-hill

Parks Victoria. (2012, December 5). *Tower Hill Wildlife Reserve*. Retrieved December 5, 2012, from Park Web: http://parkweb.vic.gov.au/explore/parks/tower-hill-w.r

Peron, F. (1809). *A voyage of discovery to the southern hemisphere, performed by order of the Emporer Napoleon during the years 1801, 1802, 1803 and 1804, Vol. 1*. London: Richard Phillips.

Ryllis Clark , M. (1996). *Discover Historic Victoria–60 heritage places to visit*. Ringwood: Penguin Books Australia.

SGS Economics and Planning,(2002). *Tower Hill Future Directions Strategy*. Retrieved December 5, 2012, from http://parkweb.vic.gov.au/__data/assets/pdf_file/0019/313444/Tower-Hill-Future-Directions-Strategy.pdf

Sydney Morning Herald, 13/6/1902

Taylor, J.E. (1886). *Our Island-Continent: A Naturalist's Holiday In Australia*. London: Society For Promoting Christian Knowledge.

Tourism Australia, (2013). *Aboriginal Tourism News*–Indigenous Tourism Champions – Tourism Industry News – December 2013, 5th quarterly newsletter. Sydney: Tourism Australia.

The Argus, 16/9/1865, 11/7/1882, 13/12/1884, 12/8/1891, 18/6/1938, 5/5/1953

The Australian Sketcher with Pen and Pencil, 1/5/1877

The Ballarat Star, 17/12/1869

The Moyne Gazette, 18/10/2012

Visit Warrnambool:*Things to do/Tower Hill*. (2012). Retrieved Deccember 5, 2012, from http://visitwarrnambool.com.au/things-to-do/tower-hill/#.UL6qMobw-o0

Fig. 8.1: Tower Hill
Artist: Eugen von GUERARD (b. Austria 1811, arrives Australia 1852, d. England 1901)
Medium: oil on canvas
Credit line: Image courtesy of Warrnambool Art Gallery, photograph by John Brash.
On loan to Warrnambool Art Gallery from the Department of Sustainability and Environment. Gift of Miss Effie Thornton 1966.

Fig. 8.2: Tower Hill Lake, Koroit, A.C. Cooke, wood engraving, *Illustrated Australian News* 21/3/1883.

Fig. 8.3: Worn Gundidj Visitor Centre. Source: Lisa Justin 9 December 2012.

Fig. 8.4: Present view from where von Guerard painted 'Outlook' Source: Lisa Justin 9 December 2012.

9 The You Yangs Regional Park By Ever Dolce and Ian D. Clark

The You Yangs Regional Park, 60km SSW of Melbourne on the Werribee Plains near present-day Lara, has been a place of importance in Victoria, long before permanent European settlement. Matthew Flinders made a trek to its summit in 1802 thus making it one of the first places in Victoria visited by Europeans, and as well as these historical values it has been a significant site of Indigenous habitation for thousands of years. The ranges themselves are much older, being from an 'isolated Devonian granite intrusion in older Sedimentary bedrock' (Waarden, 1986: 1). Webb and Marshall (2002: 6) state that the high granite hills of the You Yangs formed about 416 million years ago. During the Tertiary period, about 30-70 million years ago, 'shallow skeletal soils and uniform sand surface developed'. Erosion from the high granite hills was responsible for the forming of the foothills, during the Pleistocene period, 1.8 million to 10,000 years ago.

Since first occupation by European squatters in 1835, the You Yangs have been managed by a number of different individuals and government agencies for a variety of different resource uses. The history of this management and the evolution of the use of the park will be discussed in this chapter. The two archaeological surveys (Waarden, 1986; Webb & Marshall, 2002) conducted at the park, will be elaborated upon to demonstrate the park's Indigenous values.

Following the example set by Clark (2002) this chapter aims to understand the history of tourist visitation at the You Yangs through the use of the theoretical models of MacCannell (1976), Butler (1980), and Gunn (1994). The You Yangs is a very different entity to that of a single mountain peak, a rock art site, a waterfall, or a cave –and the implications of this for the applicability of Gunn's spatial model merit discussion. Indeed, the You Yangs range is more than the peaks and foothills that form the Regional Park – it includes the outcrop that is now a recreational area called Fairy Park near Anakie. The You Yangs Regional Park comprises only 35 per cent of the You Yang ranges, thus it is an interesting study of a tourism attraction that involves only a portion of the natural feature that it comprises. Gunn's (1994) nucleus for this attraction would initially be Flinders Peak. However, because the You Yangs offer a number of varying and unique product segments, the current boundary of the nucleus could be extended to include Acre Rock, the Bunjil geoglyph, turntable picnic area, and possibly the entire East-West walking track. An expanded nucleus could include the very popular mountain bike tracks to the west and north, and the western plantation, which is set aside for horse riding.

The You Yangs are situated within the country of the Wathawurrung Aboriginal people who have maintained a long connection with them.[34] Clark (1990) has reconstructed two Wathawurrung clans in the vicinity of the You Yangs: the Yawangi baluk centred on the You Yangs, and the Worinyaluk baluk on the west side of the Little River (Worinyaluk). Pescott (1995) argues that the Yawangi and the Worinyaluk baluk are the same people; however the demographic information available from the 1840s (see Clark 1990), confirms that they were separate clans.

Two recent archaeological surveys conducted within the park have identified at least 334 sites of Yawangi baluk habitation (Waarden, 1986; Webb & Marshall, 2002). Webb and Marshall (2002) indicate that further artefact discovery is likely, as large sections of the park are yet to be thoroughly investigated. Also, as the boundary of the park sits on less than 35 per cent of the You Yang ranges, it is likely that substantial artefact scatters exist outside of the park boundaries. The recognition of the park's Indigenous sites and value is for the most part poorly communicated, as shall be discussed below.

Today the park is renowned for mountain biking, having two biking areas with over 50 km of trails. It also has several picnic areas; scenic drives; a visitor centre; and five walking tracks including one to Flinders Peak Lookout – the highest point of the You Yangs. A permanent orienteering course has been established in the park and rock-climbing and abseiling are permitted in certain areas, as is horse riding in the 'Western Plantation'. Pescott (1995: 4) noted that the You Yangs remains a popular tourist destination with the regional park attracting some 250,000 visitors annually.

9.1 First Phase: Site Sacralization and Naming 1802–1912

MacCannell (1976) postulates that before a natural attraction can be classified as a tourist destination it must go through four phases. The initial phase of naming a location is also identified by Butler (1980) in his 'exploration stage'; where a site is named and thus becomes identifiable.

Numerous Aboriginal names have been recorded for the mountain range, along with particular peaks: You Yangs is a variant of Yawang. Other variant spellings include Yaughon (Todd Jnl, 1835) and Youang (Surveyor General in Smyth, 1878 Vol. 2: 198). It is possible that this word means 'big or large hill' or 'bald' or simply 'hill' (see Clark & Heydon, 2002); although Robinson recorded 'yow.wong' as the Woiwurrung word for 'granite stone'. This latter gloss has some credibility as the You Yangs is a series of granite ranges. Pescott (1995: 72) has offered the spurious gloss that it means 'chucking mountain or volcano' and considers that this is an odd piece of Aboriginal

34 Wathawurrung, Wada Wurrung and Wathaurong are variant spellings for the same Aboriginal language name (Clark, 1990).

lore as the You Yangs never erupted. Another unsupported gloss is 'mountain in the middle of the plain' (*The Age*, 7/4/2012).

In terms of specific peaks within the You Yangs, the highest peak, Station Peak (or Flinders Peak or Wooloomanata), was known to the Wathawurrung as Wurdi-yawang, meaning 'large hill' (Billy in Tyers Jnl in Clark & Heydon, 2002; *The Argus*, 23/10/1936). A Woiwurrung name is recorded for this peak: Darooit (Berook aka Gellibrand in McCrae, 1934:195). A smaller mountain in the larger range, Mt Anakie, was known as 'Nganaki Yawa', meaning 'small hill', although Massola (1968:7) has translated this as 'twin hills'.

The earliest European name conferred on the You Yangs was 'Station Peak' on 1 May 1802 when Matthew Flinders and three of his crew trekked to its highest peak from Corio bay. Upon reaching the peak with the highest elevation (348 meters above sea level) he named it 'Station Peak' and mapped the western coast of Port Phillip Bay.[35] Flinders (1814) sought out this peak 'to obtain from thence such a view of the upper parts of the port as would render the coasting unnecessary.' In 1824, the explorers Hamilton Hume and William Hovell named the You Yangs 'Mt Wollstonecraft' and Station Peak 'Wooloomanata'. The latter name survives as the name of a homestead near the town of Lara. Pescott (1995: 72) noted that this placename has been translated to mean 'native shelters that are clearly seen', but Massola (1968) and Blake (1977) interpret this as 'willam-gnutter' meaning 'no one lives there'. John Batman named Flinders Peak 'Mount Collicot' (Labilliere, 1878 v. 2: 94). Other names for the You Yangs include 'Wedges Range' (Labilliere, 1878 v. 2: 94). Wedge's name for the You Yangs was 'Anikai Yowwham' (Wynd, 1981: 13).

In Flinders' 1802 journey to the You Yangs he made a number of observations about the land leading up to the peaks. Being the beginning of May, Flinders (1814) details how he had difficulty finding a dry place to camp and notes that the: 'low plain where water appeared frequently to lodge; it was covered with small-blade grass, but almost destitute of wood, and the soil was clayey and shallow. One or two miles before arriving at the feet of the hills [You Yangs] we entered a wood where an emu and a kangaroo were seen at a distance'. When he had completed the survey of the bay he 'left the ship's name on a scroll of paper, deposited in a small pile of stones upon the top of the peak'.[36]

---

35 Station Peak was officially changed to Flinders Peak in 1912, via notification in the Victorian Government Gazette (Scott, 2001); presumably to coincide with celebrations of Flinders' visit (see below).

36 In 1912 the Victorian government unveiled a bronze plaque set in rock near the summit commemorating Flinders' visit. In March 1974 a plaque commemorating the bicentenary of his birth was unveiled.

Surveyor Charles Grimes came to Port Phillip in February 1803 with a mandate to survey the land for use and possible settlement. Unfortunately his account of the You Yangs is austere, only referring to them as 'the mountains'.

Hamilton Hume and William Hovell conducted the first overland expedition from New South Wales, to Port Phillip in late 1824. Their account of the journey provided John Batman and future settlers with a highly positive view of Port Phillip's capacity to support agriculture. At Port Phillip, Hovell gives us an account of the You Yangs, Wathawurrung people and the Plains of Iramoo (Werribee Plains), that Batman would later wax lyrical about (Scott, 1966).

> From the Arndell up to Kennedy's Creek [Hovells Creek] is the Duke of Clarence's Downs. This creek comes round the west end of Mount Wollstonecraft [You Yangs]. I gave it that name for the respect I have for the gentleman bearing its name. It is a very conspicuous mountain, or rather it consists of a number of peaked hills each other, particularly when it bears N. 50 deg. W. (distance about five miles).

Hovell's observation that 'to the very top of Mount Wollstonecraft is fine green grass' points to the absence of trees in the time prior to European settlement. Because the You Yangs is presently covered on all sides by thick vegetation, we can conclude that the existing 'natural' state of the park is quite different to what it would have been like in the early 1800s. It also reinforces the polysemous meaning of Yawang as 'bald'.

John Batman has generally been credited with being one of the principal founders of Melbourne in 1835 (Billot, 1979; Harcourt, 2001). Hume and Batman had been childhood friends and there is a good deal of evidence to suggest that in 1835 Batman had access to the Hume 1824 map (Billot, 1979; Harcourt, 2001; Boyce, 2011). Batman, on seeing the You Yangs, bestowed the names of three esteemed colleagues upon its highest peaks–Collicott, Cottrell and Connelly. Looking across to Mount Anakie he decided to call this Mount Solomon, after another acquaintance. Like most of the names Batman gave to Victorian places none of these names were to stick. Before leaving the area, for the Yarra River, Batman made the following observation: 'we saw the smoke of the Natives under Mount Collicott' (Flinders Peak).

The You Yangs formed part of the land claimed by John Batman's Port Phillip Association in 1835. The range fell between two blocks, one given to James Simpson, and the other Michael Connolly. However, Governor Bourke did not support Batman's land treaty and the land claimed by the association was opened up. Pescott (1995: 35) suggests that at this time, a 'section of land, including the main granite peaks, was retained as a Common and used for cattle-grazing and timber cutting'. This Common was later to become a Timber Reserve in 1866, and eventually the You Yangs Regional Park in 1992.

W.J.T. Clarke occupied the You Yangs in 1837 and asserted he was the first European occupier of Station Peak and a portion of the Little River (Bride, 1898). As one of the first European squatters to come to the area, Clarke was witness to the rapid population increase of settlers. Clarke would later become one of Australia's wealthiest people, as a result of a variety of speculative investments, and as such, was not personally engaged in the hands-on aspect of his livestock enterprise at the You Yangs (Clarke, 1980). After a couple of years at the You Yangs the run was abandoned for lands further into the interior of the colony. Clarke reports that they had no conflict with the Wathawurrung Aboriginal people (Bride, 1898; Clarke, 1980; Clark, 1995). In the 1840s, brothers Albert, Marcus, and George Synnot were significant landholders in the You Yangs.

Surveyor Charles Tyers ascended Station Peak on 14 October 1840 and from there 'obtained a round of angles, true bearings, and the latitude by circum-meridian altitudes of the sun' (*The Sydney Herald*, 1/10/1840).

By the mid-1850s the major land holders in the vicinity of the You Yangs included the Synnot brothers, John von Stieglitz at 'Station Peak', Frederick William Armytage at Wooloomanata, and Robert Chirnside at Mt Rothwell.

The earliest reference to the name 'You Yangs' appears to be a notice in the *Geelong Advertiser* (1/2/1851) alerting its readers to a mare that had strayed from the River Ex and was last seen in 'Mr Bates's paddock, You Yangs' and offering a reward for its recovery. Bates station was 'Station Peak'.

In November 1853 Government Botanist Ferdinand von Mueller undertook a botanical collecting trip to the You Yangs. In December 1853 *The Geelong Advertiser and Intelligencer* (12/12/1853) published an article entitled 'Station Peak' in which they noted that this prominent and interesting hill 'is rarely visited' and estimated that 'perhaps not fifty people have stood atop the Granite Peak'. From the summit 'a glorious scene bursts forth' of unlimited varied landscape.

In 1856 a rail link from Geelong to Little River was established and Pescott has noted how tourism leapt ahead and local towns, Duck Ponds (Lara) and Little River, vied for the trade from the many visitors to the ranges. With the advent of the rail line the You Yangs became a favourite destination for day excursionists who having travelled by train to Little River would walk across the grasslands to the mountain range. *The Argus* (22/11/1856), for example, gives an account of some 500 people who took the train to Duck Ponds and availed themselves of the fine weather to walk from there to the summit of Station Peak. 'Among these were many ladies, who displayed a degree of courage and perseverance which did them credit. The ascent on foot is not by any means an easy or pleasant task and there was many a 'spell' required before the holiday tourists reached the summit. Once there, however, they were all well rewarded for the exertions they had undergone'.

Some of the early tourists to the You Yangs have left accounts of their experiences. For example, in 1857, Louis L. Odgers published an anonymous 47-page account entitled 'You Yang's Summit: a narrative of the proceedings of a pic-nic party ...' The author considered the summit afforded a view of 'brilliant scenery, magnificent landscapes, and unparalleled grandeur of the surrounding country'. *The Argus* 28/10/1857 equated the summit of the You Yangs with the Lal Lal Falls, as objects of interest to the tourist. A correspondent in *The Argus* (30/12/1857), was glowing of the You Yangs, and stated that they are 'one of the most delightful spots for a picnic that can be found on the face of the globe':

> In these holiday times, when parties are often hard up for a locality whither to proceed to enjoy themselves, perhaps you will allow me to notice one of the most delightful spots for a pic-nic that can be found on the face of the globe; and as it is easily approachable from either Melbourne or Geelong, I have no doubt that many, when they know the magnificent prospect that awaits their eyes on reaching "You Yang's Summit," will sacrifice a little comfort, and put up with the inconvenience of climbing a steep hill, some twelve or fourteen hundred feet high from the starting point, for the sake of enjoying one of the grandest sights to be seen in the world.
>
> The highest point of the You Yang range of hills, known as Station Peak, is situate on the most southerly of that group, or the one nearest to Geelong. An excellent road, either for carriages, gigs, or horses, leads to a grassy flat between this hill and the one immediately adjoining it on the north; and in this flat there is a never-failing spring of excellent water: there is therefore no occasion to burden a party with the carriage of this necessary element. The usual "camping place" of pic-nic parties is at the north base of the highest cone a little above the spring, and close beside a cluster of prickly acacias.
>
> The ascent is by no means easy, particularly on a hot day, but with a little exertion may be accomplished by gentlemen in half an-hour. Arrived at the top, the first thing that presents itself is an enormous block of granite, almost flat on the top, and on which some tourists have erected a kind of flagstaff, supported by a large heap of small stones, piled round its base to the height of five or six feet. As there is no timber to impede the view, the eye has full range-if the weather be not hazy.
>
> The scene that presents itself southward is truly beautiful. ... But it is not alone to the rich treat of enjoying such a scene that the pleasure-seeker is limited: there are innumerable things to engage his attention as a naturalist-rare specimens of insects, varieties of trees, countless shrubs and wild flowers, and a small share of game, such as wild pigeons, quail, and snipe, are to be found about the hills. The singular and fantastic blocks of granite standing in all shapes, as though the first peal of thunder would shake them from their uneasy-looking testing-places into the vacancy below, also afford a line field for the exercise of the imagination of the picnician. On a clear day the finest and most picturesque of the country lies at the tourist's feet, as it were.

In 1856 all forested areas that were Crown Land were vested in the Department of Lands and Survey. By the early 1860s the You Yangs was experiencing considerable environmental pressure from extensive clearing of dead and green timber. In 1862 the Duffy Act allowed for farmers' commons to be created, and in February 1866 the You Yangs became a Crown Lands Timber Reserve of about 1,000 ha.

9.2 Second Phase: Framing and Elevation 1866–1992

In 1867, to coincide with the arrival of the Duke of Edinburgh, the Shire Council of Corio held a meeting at Duck Ponds (Lara). To celebrate this event a giant bonfire was planned, to be coordinated with bonfires being lit in 'other mountainous districts'. It was expected that these blazes would be visible from Gippsland to Ararat. The You Yangs bonfire was made from 100 tons of green timber. Because Station Peak was the site of a Government observatory the next highest point was selected for the site of the bonfire.

In 1868 Henry Thomas, using the pseudonym of 'Ollie' published *An Excursionists' Guide-Book* for excursionists from Melbourne. In the guide he included the You Yangs as one of many places that are within easy reach of Melbourne pleasure-seekers, and 'yet only a very small proportion of our city population have ever enjoyed their beauties' (*The Argus*, 26/10/1868).

In 1868 a Crown Lands ranger was appointed to police the restriction on greenwood cutting. The 1869 Land Act set aside 300 ha as a firewood reserve and nine 'selections' for grazing on the 'You Yangs common' were permitted (Pescott, 1995: 78). On 1 April 1869 a restriction, of 10 head of cattle per individual, was imposed upon grazers wishing to use the You Yangs to depasture their animals. According to *The Bacchus Marsh Express* (6/3/1869), as a result of '5,000 acres of the You Yangs forest' being 'set apart for the preservation of timber' in 1868, the livelihood of local fence posters who used this forest was affected. The *Express* informs us that the local fence posters had found a way around the prohibition of the felling of live trees. 'We're not prohibited from cutting down dead trees. We'll ring a lot of living ones, they'll soon die, and then we can cut them down with impunity.'

The Ballarat Star (4/9/1869) gives some detail into the extent of deforestation that had occurred at the You Yangs. Orders had been issued for police to prosecute 'any person cutting timber on the mount'. This article is interesting as it gives an account of Frederick Armytage going to 'considerable expense this year in tree-planting, and although we know several landholders who smile at his hobby, the time will perhaps come when he will have plenty of feed for his sheep, when their runs will hardly be able to furnish a blade of grass, unless they also assist nature in the manner that he has been, and is still doing'.

In 1875 the Crown Land reserve was increased to its present size of 2,000 ha, and a program of reforestation commenced.

The Illustrated Australian News (21/2/1877) gives an interesting account of the Werribee Plains which it described as 'an interminable succession of sand, pasture and sheep; sheep, sand and pasture. About midway through this setting of sun parched, arid wilderness, rises a jewel of heaped up rock and dense vegetation, the group of ten or twelve hills known as the You Yangs. To artistic wanderers they are a perfect epitome of grand effect' (see Fig.9.1).

In January 1879 the *Illustrated Australian News* published another article on the You Yangs and recommended it as a suitable location for lovers of wild nature and those who enjoy a good scramble. The article also refers to the habit of visitors to graffiti the rocks by scratching their initials or carving their names (see Fig.9.2):

> To all lovers of truly wild nature and enjoyers of a good scramble, we recommend this route to enjoy a holiday. The You Yangs are a group of thickly-wooded volcanic hills between Melbourne and Geelong, about, two miles from Little River station, from which township an excellent road runs into the town common at their foot, and such a town common as for picturesque beauty and singularity has not its equal in the colony, perhaps not in the world. Huge basalt rocks, in every conceivable form that volcanic upheaval, ocean waves, rain, sun and frost can produce, are there in profusion. Some steep hills, with sides serrated with stone, and glittering with varied foliage; others glowing with heaths, and having broad tracks of close springy grass extending to the very summit. It is like a bit of pre-Adamite world, and in its every peculiarity the absolute opposite of the entire country around it. The highest peak of this group is Station Peak; its height is about 1600 feet, and it is crowned by the wooden erection used by the geodetic survey. It is a favorite resort of picnic parties from Geelong, and, occasionally, Melbourne. It is profusely scored with initials of its many visitors, and with the magnificent prospects it presents on the landward side of its outlying hills, flanked by extensive plains, and bounded on the far-off horizon by the dim line of the Anakies, and on the seaward by the Werribee Plains, the Bay and the opposite coast. It is rather puzzling what queer mental peculiarity possesses Englishmen who take the trouble to face so formidable an upward scramble, and causes them to spend time and break pen- knives in carving their names in hard wood, almost hardened into stone by exposure to all sorts of weather. The only wonder is that the easy distance from Melbourne of the You Yangs has not made them a more, popular resort for pleasure-seekers and sportsmen (*Illustrated Australian News*, 22/1/1879).

With the formation of the Field Naturalists Club of Victoria and Geelong Field Naturalists Club in 1880 regular excursions to the You Yangs were organized for their members to explore the granite ranges. Several reports of excursions in the 1880s were published in *The Victorian Naturalist* and *The Southern Science Record*. The first account presented a dismal picture of the You Yangs 'which are simply piles of immense boulders heaped one on top of another, with scant vegetation and a somewhat desolate appearance'. The writer warned that 'Water here is scarce, and tourists should provide themselves in this regard before starting. To the sportsman, this district should be a perfect Elysium, as rabbits swarm from every rock and hill; but to the collector of plants and insects, it will, I am afraid, prove but a very indifferent field' (Pescott, 1995: 107). Pescott notes that despite these unflattering comments members of the Field Naturalists returned to the You Yangs; visits took place in 1883, 1886, 1908, 1916, 1917, 1933, 1936, and 1960, for example.

The increase in timber extraction from the You Yangs in 1885 culminated in an outcry from the residents of the surrounding area; who believed that due to excessive extraction, less rainfall was the consequence (Waarden, 1986: 6). By 1896, some 480,000 trees had been planted in over 300 ha. The grazing commons were abolished in 1904 and over 1,400 ha declared state forest.

> A DAY ON THE YOU YANGS.
>
> To those whose holidays are limited to occasional days, or even less, and who have not as yet made the trip, a few hours on the You Yangs should come as a welcome break to the usual bay excursion, or suburban seaside township, with its monotonous drift of sand, and low uninterest-

ing cliffs. Scenery of whatever kind has certain charms, which may, and does appeal to lovers of nature, but the craving for what is greatest and best can only be satisfied by communion with nature amid the grandeur of her mountains. This is the experience of all who have set out to find the best the world has to give to her of times unworthy sons and daughters.

Again comes the call of the hills, and again we must respond. We would like to once more climb the highest peaks of our Australian Alps, or even wander amid the ranges of the Great Divide. Unfortunately we must leave these for some future time, but we can renew our friendship with the You Yangs, and so while the holiday trains are crowded with happy people speeding away to distant scenes, we content ourselves with a humble day-return to Little River. Never mind, we are going to a mistress, whose charms never fade, who will continue to unfold fresh and yet fresher beauties.

Just before the train enters Little River station, it crosses a road. This road takes us in a straight line to within half a mile of the foot of Wurdi-Youang or Station Peak, the highest of the Anakies or You Yangs. A gentle walk of three-quarters of an hour having brought us to an iron gate, and the end of the road, a track will be noticed branching off to the left toward the foot of the main ridge. We follow this track, keeping straight toward the lower end of the ridge until we come to a rabbit proof fence (the You Yangs are infested with rabbits). Crossing this we see a broad plough track (cut for a fire break) over a clear part of the ridge. Two or three minutes along this track, and we are on the centre of the ridge, and can now begin the ascent proper. The western face of the ridge is more rock strewn and rougher than the eastern, so in skirting boulders and obstructions, it is better to keep to the right, always getting back to the centre as soon as convenient. Ten minutes or so and we clamber the boulders forming the summit. This is the easiest path, and the one generally followed, and can be accomplished by ladies and children without much undue effort. The ascent can be made from any point by the average rock climber, though there are parts where the help of one's climbing boots and baton will be appreciated. Although the ascent has been easy, there is nothing little or ordinary about the view. Of its kind, the view from the top of Station Peak is unsurpassed in Australia; Isolated and abrupt, in the midst of a vast plain, a panorama lies at our feet that takes a more powerful pen than ours to adequately describe. In fact few, very few, have possessed the gift to give to others their own impressions of scenery, and this must of necessity be so, as nature speaks and appeals to us in way that are really beyond language. From the Queen City of the South away on our left, to the admirably situated city of Geelong on the right, one level spread of cultivated plain gives us some idea as to the extent crop growing is carried on even at our doors. Seemingly almost at our feet, though nine or ten miles away, the sea sparkles and glistens, looking about as wide as a river estuary, but it is twelve miles across it to the undulating cliffs of Portarlington. We are lucky in being favored with fine clear weather to-day. We can pick out the white sails of small craft pleasantly sailing along, and the "Courier" steaming to Geelong looms up quite large. The country to the north and west is different, in that very little cultivation can be seen. We can see, however, clear up to the Great Divide forty miles away, but except for a few miles in front, the country has the appearance of being virgin bush. As before noted, the western face of the mount is the wildest, almost approaching to desolation. The country to the west too seems wild and uninhabited, yet we confess that we like this side the best. Perhaps it is the setting sun "hallowing with golden rays." We catch ourselves thinking: Will the world be better and happier when the western plain shall have become as the eastern. Let us be charitable, and hope so. At the foot of the peak to the south west, a large rock will be noticed in the shape of an inverted basin. The rock covers about an acre and a half of ground, and is estimated to measure 4,000,000 cubic ft. It is probably the largest detached rock in Victoria. On the top-the part that corresponds with the rim of the basin-amidst the dust of ages, several trees are growing quite contentedly. It must be remembered that the cracks and crevasses of rocks hold water for a long time. This accounts for the trees on the mount being of healthy and luxuriant growth. It may also account for the fact that, on our last visit, although we encountered no grasshoppers on the plains, the hills were swarming with them. Another curious thing we found: lizards like cheese. While eating our humble lunch on the summit, a good sized-black-rock-lizard came along, and squatted in front. Probably he makes this rock his home, and is used to strangers. Anyhow he looked friendly, so we threw him a piece of bread. He greedily licked off the butter, and made a valiant attempt at the crust, lizards jaws are not made for chewing hard tack, and he was distinctly beaten. In fun, we threw him a piece of cheese, and I never want to see anything more comical than a lizard wolfing cheese (wolfing is just the word). My friend had a good feed. I hope he did not suffer unduly from indigestion.

It is not likely that the You Yangs have ever been of great height-the present height is 1,154 feet above sea level; the mount proper is 1,003 ft. high at most not more than 2,560ft., as proof of this, one can step off the plain on to the mount proper at almost any part, and more particularly opposite the highest point, facing south east, where the side is most precipitous (This is also the most-picturesque). Had the range been of any considerable height in former times the disintegration would have made the approaches more gradual. The surrounding country is basaltic. The range is of gneiss formation. I have not been able to find any natural signs of gneiss more than 500 yards from the base on the eastern side (the most precipitous). The presence of feldspar favors the theory that the range owes its formation to volcanic agency. Though why in so many instances isolated mountains should differ in basic formation from their immediate surroundings is yet unexplained, and personally, much as I would like, I am quite unable to give a feasible explanation to this curious phenomena. Few people would suppose the You Yangs as high as Mount Macedon, yet the mounts proper are within a few feet (15ft. to be exact).The country rises for miles, until the base of Macedon is reached, which gives the mount its majestic and massive appearance. Just a few words in conclusion. The best time to pay a visit is in the spring. One gets greater contrasts, and a good field glass should be taken. During spells of hot weather, haze will usually obstruct the view, should the wind be from the north. Rocks get very hot, and give out great heat. A day-or-two after rain is a good time to make an ascent. Remember that the view of a mountain may be as satisfying and good as the view from it. If you have an eye for the beautiful, with ourselves, not the least of your pleasure will be the contemplation of Station Peak from the near plain, and may be the memory that wilt linger longest of your day on the You Yangs. J.T.C. (*North Melbourne Courier and West Melbourne Advertiser*, 4/2/1910).

In September 1916, some members of the Field Naturalists Club of Victoria made an excursion to the You Yangs. The results were published in *The Victorian Naturalist* (1916: 89):

A report of the excursion to the You Yangs on Saturday, 16th September, was made by one of the leaders, Mr. R. A. Keble, who stated that a number of members and friends took part in the outing. The party left town by the 6.30 a.m. train, reaching Little River at 7.40, and immediately set out for Station Peak, the crest of which was attained about mid-day. After spending some time viewing and dealing with the physiography of the surrounding country, a leisurely descent was made, and a northerly course maintained till a small spring was reached. In the immediate vicinity of this spring a number of skinning flakes, scrapers, &c., relics of the aboriginals, were picked up. The party left the reservation about 3 p.m., returning to town by the 6 p.m. Geelong train.

In 1924 *The Argus* (25/11/1924) included the following account of the You Yangs in an article on attractions in the Geelong district: ‘North of Geelong are the You Yang

Mountains. From Flinders Peak (1,154ft.) one of several lofty points, there is a delightful outlook over many square miles of level country, with the waters of Port Phillip Bay beyond. The You Yangs may be reached by train from Melbourne to Little River, and afterwards by a walk of from five to six miles, or to the foot by motor-car from Geelong or Melbourne'. Another travel account was published in 1950:

> Back Tracks For Weekenders By PAT FRANKLYN
>
> WHEN you have had your fill of looking at photographs of rock formations in Central Australia and regretting that you cannot afford either the money or the time to visit them, spend a day in the You Yangs. Doubtless you have seen this range often enough from the beaches on the eastern side of Port Phillip Bay, looming pale blue and jagged across the water. If, on a clear day, you scramble up over the spectacular granitic rock to the top of Flinders Peak, you will see this view in reverse; and, moreover, find Port Phillip Bay spread out, at your feet.
>
> It is of interest that the soil and rock of the You Yangs are some of the oldest in Victoria, and much older geologically than the plain over which you walk from Little River. It is also of interest that Flinders Peak was climbed by some of the first Europeans to visit this part of Victoria. John Murray, sailing from Port Jackson in the Lady Nelson, discovered the entrance to Port Phillip Bay in January, 1802. Ten weeks later Matthew Flinders, in the Investigator, having sailed from England to undertake an exploration of the whole coast of the continent, independently discovered Port Phillip Bay. In the course of a rapid three-day survey of the bay, he landed with a small party on the northern shore of Corio Bay and climbed to the top of the Peak, giving it the name of "Station Peak" (*The Argus* 29/12/1950).

In 1958, 355 ha were allocated as a scenic reserve and in 1962 an additional 110 ha were given a special purpose reserve status, for the protection of birds. In 1964 the reserve was declared a forest park. Then in 1981 an additional 1515 ha were added, bringing the total size of the park to close to 2,000 ha (Waarden, 1986: 4). Finally, in 1992 the You Yangs Regional Park was gazetted. Its function was 'to provide opportunities for informal recreation for large numbers of people, and to conserve and protect the natural ecosystems to the extent that this is consistent with the recreational role' (Pescott, 1995: 79).

In 1969 Aldo Massola published a narrative of a car journey to Victoria's primary Aboriginal heritage sites. He included the following discussion of the You Yangs:

> Historically, too, the You-Yangs are interesting. In 1802, Matthew Flinders climbed the highest peak of this range (which now bears his name) in order to survey Port Phillip. He, and three of his sailors, walked to it from a point in Corio Bay opposite the Peak, a distance of over ten miles, but apart from weariness they only suffered from thirst. In his Journal, Flinders stated that neither on the way up, nor on the return journey, could they find any water. Although he did not know it, water was quite close to him on the Peak. It was in one of those strange water wells hollowed in granite, a cavity developed over the centuries, through the impurities in the hard rock being first softened and then washed away by the action of the elements. The Aborigines no doubt helped nature and accelerated the process by digging into the original fault, thus enlarging and deepening it.

> This particular basin is on the lower side of a gently sloping granite boulder, protruding over the surface of the soil. The basin measures forty-nine inches by forty-two inches and is twenty-eight inches deep. Consequently it holds rain water well into the dry season. The rock itself is easily found. It is on the south-west flank of the peak, about 200 yards from the top, and in line between the summit of the Peak and Acre Rock. On Acre Rock, that huge mass of granite said to cover one acre, there are several of these basins, but they are only shallow depressions, and hold little water. However, they must have been used by the Aborigines, since a number of microliths and other small stone implements have been picked up near them. Given time, these shallow depressions would all have become water storage basins.
>
> The Aborigines that roamed over and around the You-Yangs were the Yawangi, one of the many groups of the Wothowurong tribe, whose territory stretched as far as, and beyond, Geelong and Ballarat. Upon the plains, in the early days, there was an abundance of game and edible roots of many kinds and everywhere there was surface water. Under these conditions the Aborigines moved freely, and were able to camp almost anywhere, not being restricted to the water courses or the sea front. Like all natives, however, they preferred to camp on a sandy spot. One such spot is about three and a half miles north-west from Flinder's Peak, and about one mile north-north-west of Wooloomanata home station. It is a sandy patch, or "blow," with an area of between four and five acres, situated on the south-west of one of the low undulations so prevalent on these plains. Close to the southern and western sides of the blow there is a stretch of low-lying land which has all the appearances of once having been a swamp. In wet weather two near-by creek beds help to drain the plains and feed what are now dams but could well be the remains of this ancient swamp. These conditions would have been ideal for a camp site, and in fact, the entire surface of the blow is almost covered with chips, flakes, and broken stones of all description, mixed with hundreds of implements, thus proving the site's suitability.
>
> The Aborigines continued to live in this area even after the occupation of the land by the colonists and the building of Wooloomanata. It is known that Frederick Armytage and his wife, who then owned the station, took a kindly interest in the natives, especially in Billy Leigh, who was "king" of the Yawangi at the time. Billy was baptized and confirmed in the Church of England at Lara, and when he died, in August 1912, the Armytages erected an upright cross on his grave in the Eastern Cemetery (Massola, 1969: 11-13).

According to the Victorian Environmental Assessment Council (VEAC, 2006), the current status of You Yangs as a regional park does not necessarily give it a protected status.

> National and state parks provide the highest level of protection for natural features such as flora and fauna and landscapes, and for Aboriginal cultural sites and places and historic sites. Accordingly, harvesting of forest products, grazing by domestic stock, and hunting and firearms are normally not permitted, and national and state parks are exempt (in most circumstances) from exploration and mining under the Mineral Resources Development Act 1990 (VEAC, 2006).

The Department of Sustainability and Environment (DSE, 1996) and the Department of Primary Industries (DPI, 2000) state that the You Yangs sugar gum plantation has been harvested for a variety of commercial timber purposes. However, in an interview, conducted on 30 January 2013 with Mark Whyte, Parks Victoria ranger and team leader, he stated that the plantation is not currently managed for commercial production and harvest of the timber. He stated that management of the timber plantation would normally ensure that the timber is suitable for a variety of uses, such as poles and posts. Because this hasn't occurred, the most probable use of any such harvest would be for fire wood, thus the likelihood of any future harvest is doubtful.

Commercial quarrying has taken place at the northern border of the park and it is still visible. According to Webb and Marshall (2002), two non-active quarries are within the park's boundaries and one of these is adjacent to a commercial quarry just outside the park, at the northern end of the Kurrajong plantation. A live firing range is located on the park's western border and it is possible to hear the rifle bursts, when at the western end of the East-West walking track.

It is clear that the You Yangs would qualify as a National or State Park if not for the commercial activity that has encroached upon what Gunn's (1994) spatial model would consider to be its 'inviolate belt'. In order for the fore-mentioned commercial activity to be restricted to the park's 'zone of closure', either the boundaries of the park would need to be reduced or the land upon which these activities are taking place would need to be acquired by the State and incorporated into the park. Unfortunately, as is reported by Waarden (1986), the mining activity in the area is most likely responsible for the destruction of archaeological evidence. Thus, any reduction in the park's size would most likely further add to evidence destruction.

The earlier designation of the You Yangs as a scenic reserve in 1992, gave the park a status that corresponds to MacCannell's (1976) 'framing and elevation phase'. This also fits with Gunn's (1994) 'inviolate belt' zone, in that this reserved section of the site is expressly differentiated from the rest of the plantation. Butler's (1980) 'involvement' stage appears to have preceded the 1958 special purpose reservation, as there is evidence that a tour bus was operating from Little River in 1912.[37] The train station at Little River allowed visitors travelling from Geelong and Melbourne to utilize the service of the tour bus.

9.3 Third Phase: Enshrinement 1912–Present

MacCannell (1976) has identified 'enshrinement' as the third phase in the evolution of attractions. It is represented by a material increase in visitor numbers to an attraction that signifies the site's prominence, as a result of enhanced reputation. Due to the distance between the You Yangs from Little River and Lara, it is probable that visitation numbers to the park increased in tandem with the automobile revolution of the 20th Century. The improvement in roads and travel times, made over the last few decades, would also have been a contributing factor to increased tourism.

37 The information room at the You Yangs Parks Victoria office has a dated photograph available for viewing.

In 1914 a walking track to Flinders Peak was completed, in order to provide visitors with an easy path to the summit. *The Geelong Advertiser's* (26/5/1914) article entitled 'You Yangs track' reveals that 'a gentleman aged 84 years completed the climb from the road terminus over the new track to the summit in 40 minutes without distress'.

In the beginning of the 20th Century, mention of the 'spring of excellent water' was again made. The *Geelong Advertiser* (20/5/1915) article '*You Yangs Spring*' states that in order to protect the spring 'against contamination by animals and careless picnickers' it was fenced off. In a second article, the *Geelong Advertiser* (29/5/1915) reported that ranger John Blair was responsible for the work of protecting 'the perennial spring at the You Yangs' and that this 'is a decided improvement to the resort'. A third article (*Geelong Advertiser*, 2/7/1915) revealed that Mr 'Hennessy, instructor in chemistry at the Gordon Technical College' had conducted an analysis of the water from the spring 'with a view to determining the chemical properties of the You Yangs-spring, which is largely used by excursionists to this popular resort'.[38]

This spring was not mentioned by either Waarden (1986) or Webb and Marshall (2002). It would have been a very important feature to the Wathawurrung and its existence should call into question some of the assumptions about habitation within the You Yangs, and the proximity to cultural sites that were previously thought to have no water supply. Having said this, it is worth quoting Mark Urquhart (personal communication, 16/9/2013), ranger in charge at the You Yangs Regional Park:

> Unfortunately, none of the staff at the You Yangs are aware of the mentioned spring, including staff that started here in 1972. It is very unlikely that any fencing or remnants of fencing remains, given the catastrophic fires of 1969 and 1985 that impacted the Saddle area and wider park. The western slopes of the Saddle were also cleared and established as sugar gum plantation or utilised for gravel extraction in the early 70s. These works may have removed any established fencing. These works altered the alignment of Pannican Gully (the gully below the Saddle on the western side).
> I find a natural 'artesian' spring to be highly unlikely given the granite monolith base that the You Yangs Range is. The Granite monolith is impermeable to water and has little to no fissures. A natural spring could very well be explained by sub soil gravity fed water flow (non-artesian), which is common on the southern and eastern slopes of the range. This effect remains visible at numerous locations in the park for a number of months following saturating rains. It is caused by water (rain), permeating through the soil, hitting the base rock (granite) and following the rock down slope to a point where the rock is exposed on the side of the hill. These sites are numerous, but have no significant flow or water quality values, and definitely do not have 'spring or mineral water' taste.

Pescott (1995: 104) makes reference to the spring when talking about the 'mysterious cave'. Apparently 'it is said that in a basin-shaped mass of granite near the

---

38 The chemical analysis returned the following result: total solids, 91 grains per gallon; silica, 1.4; iron oxide, .81: calcium sulphate, 3.07; calcium chloride, 6.46; magnesium sulphate, 20.00; magnesium chloride, 8.36; sodium chloride, 50.90.

summit of the giant peak there is a spring of water'. It is interesting that *The Argus* (30/12/1857) confidently stated that owing to the water that can be found at the spring 'there is therefore no occasion to burden a party with the carriage of this necessary element'.

The tourism centre of activity in the You Yangs has shifted over the course of its visitation history. From the late 19th century until sometime in the 20th century the Saddle, to the north of Flinders Peak, was consistently mentioned in visitation reports, as the picnic area. Today, the main entrance, the Big Rock car park and the Flinders Peak car park are the main focal points of the park. Picnic areas with barbeque pits lie in the low areas to the south of Flinders peak.

The *Werribee Shire Banner's* (27/7/1925) 'Lara' reports that a scheme for the improvement of the You Yangs as a tourist resort had been initiated by interested parties from around the Corio area. A You Yangs Improvement Committee had been established in order to improve the roads within the 'forest'. The existing graded road to the Saddle and the footpath from the Saddle to the Summit would be improved. Additionally, a rotunda would be erected at the Saddle.

Applying Butler's (1980) life cycle model, a 'decline' in visitation numbers during the first decade of the 21st Century was followed by 'rejuvenation', in the form of additional funding that has resulted in visitor numbers returning to their historic 1981/82 high of 250,000 visitor days. The dropping of the park's entry fee would also have been a contributing factor. Enquiries made to Parks Victoria revealed that visitor counts have not been systematically conducted over the years that the various government agencies have managed the You Yangs. The few counts that were conducted were done using statistical sampling techniques (M. Whyte, personal communication, 30/1/2013). The Land Conservation Council of Victoria (1985, as cited by Waarden, 1986: 58) states that the You Yangs achieved its highest recorded visitations in the year 1981/82. This was again matched in 2012 with 250,000 visitors (Parks Victoria, 2012). However, in the interim years the visitation counts conducted show a dramatic decrease in visitors. In the four years from 1997/98 to 2000/01 visitation counts average 83,069 (Parks Victoria, 2013). Park entry fees were abolished as of 1 July 2010 and this could explain the divergence in visitor counts (Parks Victoria, 2010). Interestingly, it is noted that an increase from 5,000 to 100,000 mountain bike riders occurred in 2012 (Parks Victoria). This would suggest a demographic change in demand for the park, possibly as a result of funding being provided for extensive mountain bike tracks. Whyte (personal communication, 30/1/2013) stated that current counts are conducted at various track locations and that the results of these counts were used to deliver the 2012 count, of 250,000 visitor days.

Because the You Yangs is not a State or National Park, Parks Victoria is not obligated to produce management plans (VEAC, 2006). Such plans generally provide insight into management's decision making, based upon conditions prevalent at the time of the report, and can include details about park usage and visitor counts.

9.4 Fourth Phase: Duplication 1847–Present

MacCannell's (1976) fourth phase in the development of attractions is that of 'duplication', when copies of the nucleus of the attraction, in this case the ranges, are made available through media such as paintings, photographs, and postcards. As early as 1847, the You Yangs began to be subject to 'duplication' as a consequence of its significance to the settlers of Victoria. One of the first paintings featuring the You Yangs is by John Skinner Prout, painted in 1847, and although this painting focuses on Geelong, the You Yangs are clearly visible in the background. Robert Hoddle's 1847 watercolour 'Station Peak from St Kilda, 1847' is another painting from this period that presents a view of the You Yangs in the distance. 'View of Station Peak' by George Alexander Gilbert, was painted in 1856, brings the You Yangs into focus and is probably the first art work to do so (Fig.9.3). Emma von Stieglitz in 1854 painted various views of the You Yangs, including Station Peak, and a deserted Aboriginal mia-mia at Villamanatta (see Pescott, 1995: 20). In 1857, Horace Burkitt, a postal clerk stationed temporarily at Geelong, painted two small water-colours of the You Yangs (see von Stieglitz, 1964). Fred Williams, a prolific Victorian artist, also created a number of works depicting the You Yangs in the 1960s. The You Yangs have also been photographed by many noted photographers, including the geologist Richard Daintree in 1861; in 1866 Eugen de Balk included a photograph in his *Album of Geelong Views*; Fred Kruger photographed the ranges in the 1880s. Samuel Calvert's (1885) 'Scene in the You Yangs' is interesting as it depicts cattle being present after the 1866 act of turning the park into a timber reserve (Fig.9.4). This is corroborated by *The Bacchus Marsh Express* (6/3/1869) article mentioning depasturing restrictions at that time.

The above mentioned artworks attest to the use of the You Yangs by artists over the decades since colonisation. However, it is the silhouette of the entire park that is the main focus of such art works, due to their recognisability. The nucleus, which includes Flinders Peak, is a contributing part to the park's recognisable vista. This differs from the example of Lal Lal Falls where the nucleus, being the falls themselves, was the sole subject of duplication. MacCannell's (1976) 'duplication' phase is realised as the park has evidently been used as a subject of numerous representations since its 'discovery' by Europeans.

In a satirically posed and highly negative review of *You Yang's Summit* (Odgers, 1857) *The Ballarat Star* (27/11/1858) goes to some length to mock the work, which they erroneously attribute to Stitt Jenkins, a bricklayer by trade and a leading temperance advocate and later prominent figure in civic affairs. This work is the first detailed writing that has the You Yangs as its backdrop, although it deals primarily with the characters and their drama while hiking. *The Star's* main objection to the work is that it focuses on the folly of these shallow people when so much could have been written about the hardships of the people of the area surrounding the You Yangs.

9.5 Fifth Phase: Social Reproduction

For MacCannell, the final stage of sight sacralization is social reproduction which 'occurs when groups, cities, and regions begin to name themselves after famous attractions'. In the case of the You Yangs, social reproduction has occurred in the naming of the You Yangs Hotel. Other local businesses include You Yangs Medical Clinic and You Yangs Yoga Centre, both in Lara. The Melbourne steam ship company purchased the steamer Kieff in 1863 from London and renamed it the You Yangs; subsequently bringing it to Melbourne for service between Melbourne, Sydney, and Newcastle NSW; it was wrecked on 12 June 1890 on a reef near Pelorous Island, South Australia.

9.6 Indigenous Values at the You Yangs

Unfortunately, the Indigenous significance of the You Yangs has not been adequately communicated to the wider community. Nora van Waarden (1986) identified 45 sites within the park that facilitated habitation and/or ritual, but noted that due to the passage of time, site disturbance and the fragmenting of Wathawurrung cultural knowledge, it is unlikely that we shall ever be able to answer questions that arise as to the utility, and significance of the park. The recent survey by Webb and Marshall (2002) increased the count of identified sites by 289.

The high elevations of the peaks offered the Yawangi an obvious advantage in times of conflict; this being the ability to survey the land in order to spot an approaching enemy. The Werribee River to the NE was a boundary between three tribes, the Wathawurrung, Woiwurrung and the Boonwurrung (Clark, 1990). As signalling was an observed way in which Indigenous people would communicate an approaching danger (see Learmonth, 1853; as reported by Bride, 1898: 40), it is likely that the three Wathawurrung clans of the area gained some strategic advantage from the panoramic opportunities offered by the You Yangs.

Waarden's (1986) archaeological report details the presence of the Yawangi baluk over much of the park's area. The You Yangs has four rock well sites that were used by the Yawangi. At each rock well site, evidence has been found to show that habitation occurred there. Situated in an unpublished location, north of the You Yangs and close to the Little River, is an arrangement of about 100 stones that have been given the name Wurdi Youang. Norris (2010) states that this stone arrangement is a significant example of Indigenous knowledge of astronomy prior to European contact. He details how the arrangement 'indicates the setting position of the sun at the solstices and at the Equinox'. In an interview conducted on 30 January 2013, Wathawurrung elder David Tournier, Cultural Heritage Education Language Officer Wathaurong Aboriginal Co-operative, at Geelong, explained that the three large stones at the western end of the arrangement align with the Anakie Hills to measure the movement of the sun.

Tournier (personal communication, 30/1/2013) said that there used to be a representation of Bunjil, with its origin in prehistory, located at the You Yangs. He states that the representation had been situated close to the current Flinders Peak car park, before being 'destroyed'. A photograph, provided by Tournier, depicts a very large stone in a shape that resembles a beak, with the caption 'this rock was remove[sic] from the you-yangs [a] long time ago (this rock is Bun Jil's beak)'. According to Tournier, a large cover stone once existed and covered the Big Rock's main water hole. This stone's function was to ensure that clean water was available all year and acted to reduce evaporation, and promote condensation. He spoke about an event that he had witnessed when he was 8 years old. He states that at the time, a number of older boys pushed the cover stone off the water hole and over the side of the big rock, where it shattered on the ground below. He also stated that the Big Rock was 'traditionally' used as a 'school' for the education of young Wathawurrung people.

As part of the Victorian Aboriginal Heritage Act of 2006 the Wathaurung Aboriginal Corporation, based in Ballarat, is the registered Aboriginal party responsible for decision making in matters of Indigenous cultural sensitivity at the You Yangs (DPCD, 2012). Prior to the 2006 Act, the Wathaurung Aboriginal Cooperative, based in Geelong, was the responsible party.

9.7 Tourism at the You Yangs

The facilities at the You Yangs are scattered around the park, with rest rooms located at the park's entrance, the Big Rock car park and the Flinders Peak car park. These are modern facilities that provide running water, toilet paper and appear to be regularly serviced. Barbeque and fire places are situated throughout the lower elevations, and at both of the fore mentioned car parks. Orienteering, mountain bike riding, bushwalking, abseiling and horse riding are activities that can be experienced while at the park. Numerous information boards are evident at various locations. The content of these boards provide visitors with basic information regarding the geology, history, flora and fauna, activities, and an Indigenous perspective. However, this perspective is focused on the last known living Yawangi person, King Billy Leigh, and does not elaborate upon the Yawangi baluk clan or Wathawurrung tribe in sufficient detail.

At the beginning of the Flinders Peak walking track and the East-West walking track, there are two smaller information boards that narrate the experiences of Matthew Flinders and William Buckley. It is unfortunate that a greater representation of Yawangi and Wathawurrung culture has not been made available to visitors of the park. It is reported that the initial European contact with the Yawangi included a friendship between the squatter Robert William von Stieglitz and the then Yawangi clan-head Murrydenneek (Clark, 1998). The Museum of Victoria is in possession of a large number of Yawangi artefacts in its Sutton collection, sourced from an area near the Wooloomanata Station (Waarden, 1986: 17). Perhaps at some point, part of this collection will be on display at the park, thus shining some light on the Yawangi people's presence at the You Yangs.

As part of the Commonwealth Games hosted by Melbourne in 2006, Andrew Rogers was commissioned to create an artistic representation of the Kulin creator Bunjil. This was done through the medium of a geoglyph. This geoglyph is visible from the high elevations of the You Yangs East-West walking track and from Flinders Peak. Visitors are able to walk to and amongst the stones that make up this artwork.

Once renowned as a picnic destination, today the You Yangs attracts a very different type of tourist. A possible reason for the shift in focus from the Saddle to the current areas of high use could be explained by the change in use of the park and the change in demand from picnicking to other forms of recreation. Bike tracks crisscross large tracts of the park and many of the large rock outcrops have metal ring attachments to assist abseilers.

9.8 Management Issues

According to Pescott (1995), the Field Naturalist Club of Victoria provided its first report on the You Yangs in 1881; however, they paint 'a rather dismal picture of the You Yangs'.

Since the introduction of boneseed into the park in the 1940s, as a means to reduce erosion, this plant has posed a major challenge for the conservation and viability of native flora and fauna within the park (Roberts, 2008). Today a number of stakeholders actively participate in the eradication of this aggressive plant. This eradication program is supported by schools, volunteers and the park management. Roberts has detailed the various methods employed to eliminate this invasive plant and an active burn back program is underway; as this has been proven to be the most effective means of controlling boneseed.

As a result of three flood events, the park closed in February 2011. State government funding of $1.5 Million was given to commence recovery works and in August 2011 the park was reopened to the public (*Geelong Advertiser*, 8/8/2011). The main reason for the large budget was that the floods had uncovered asbestos culverts (pipes) at 90 locations around the park. Subsequently, 80 tons of asbestos were removed and 100 pipes replaced (*Geelong Advertiser*, 1/3/2011; Pearson, 2011). The park has also had a number of severe fire events. The most dramatic occurred on 8/1/1969 and burnt the entire park, and destroyed much of Lara and resulted in the deaths of 17 motorists on the Geelong to Melbourne freeway (DSE, 2012).

Waarden (1986) states that because of abundant vegetation cover and inaccessibility, many of the rock-wells and scatter sites are not in danger of being discovered or damaged by visitor use. However, a number of scatter sites are located in areas that have a heavy visitor presence and as such require continued monitoring. A number of recommendations that were made by Waarden (1986) have not been implemented.

These recommendations centre on providing educational material and information to the public and in particular school groups. Waarden emphasised providing material that could be handled; as well as display cases that could house local artefacts. At the time of the survey (1986) they were stored at the Museum of Victoria. The current information room, located at the park's office, does have information about Australian Aboriginal culture; however, this is largely focused on generic information that is not specifically focused on the Wathawurrung tribe.

A second archaeological survey was conducted in 2002 by TerraCulture Pty Ltd, on behalf of Parks Victoria (Webb & Marshall 2002). The aim of this second survey had two stages. Stage 1 was to document the archaeological evidence in the areas of the park that Waarden (1986) had not included in her 1986 survey. This included the lowland areas of the southeast and western part of the park, and the uplands to the northeast (Webb & Marshall, 2002). Stage 2 was to provide Parks Victoria with a risk assessment analysis and recommendations for management of the park. These recommendations were generally designed to provide rangers with a protocol to follow in the event of direct contact with archaeological material.

9.9 Conclusion

The You Yangs have been shown to be a different entity to the sites that are the subject of the other case studies in this volume. It is not a single waterfall or a single mountain but rather a range of granitic hills, of which approximately only a third have been formed into the You Yangs Regional Park. The largest peak in the You Yangs is one of the earliest sites in Victoria that were visited by a European – in this case Matthew Flinders in 1802.

In 1835 part of the You Yangs was retained as a Common and used for cattle grazing and timber cutting. This became a Crown Lands Timber Reserve of about 1,000 ha in 1866. In 1875 the reserve was increased to its present size of 2,000 ha. In 1958 some 355 ha were declared a scenic reserve, and in 1962 an additional 110 ha was given special purpose reserve status for the protection of birds. In 1964 the reserve became a state forest. In 1981 some 1,515 ha were added bringing the park to 2,000 ha. In 1992, the reserve became the You Yangs Regional Park –to provide opportunities for recreation and to protect and conserve natural ecosystems to the extent that this was consistent with its recreational role.

It has been a site for botanical research since 1853 when the government botanist von Mueller conducted field surveys there, and with the formation of the Field Naturalists Club of Victoria in 1880, regular excursions have been conducted to the reserve. In 1853, the You Yangs was not the subject of much visitation, and the *Geelong Advertiser* speculated that perhaps as few as 50 people had stood on its highest peak. The advent of the railway from Geelong to Little River saw the You Yangs become a favourite destination for day excursionists from Melbourne and Geelong.

The You Yangs has been and remains an important historical attraction, especially for people who are interested in the early colonial period when Europeans first explored and settled in Victoria. There is a wide range of diverse activities available at the park. The recent surge in mountain bike rider participation attests to management's successful efforts in rejuvenating visitor interest. The viewpoints of Butler (1980), Gunn (1994) and MacCannell (1976), have been used to show the change in use and visitation at the You Yangs. Today the You Yangs offer visitors a variety of experiences, through the provision of modern amenities and infrastructure. The changing demographic demand for the park and the continued involvement of the two Wathawurrung organisations, allows for an expanded offering of Wathawurrung culture, and history. Thanks to the archaeological work carried out by Waarden (1986) and Webb and Marshall (2002), evidence is available to show the Indigenous connection to the You Yangs. Should certain recommendations of Waarden (1986) eventually materialise, the public could have access to a much richer experience than is currently offered at the park. The Wathaurong Aboriginal Cooperative could perhaps take advantage of the infrastructure already available, such as the Duck Ponds school building; in order to conduct educational and welcome to country interactions, with schools and other groups.

References

Balk, E. de (1866). *Geelong Album photographed by E. De Balk*. Geelong: de Balk.

Billot, C.P. (1979). *John Batman: The Story of John Batman and the Founding of Melbourne*. South Yarra: Hyland House.

Blake, L.B.J. (1977). *Place Names of Victoria*. Adelaide: Rigby Ltd.

Bride, T.F. (Ed.). (1898). *Letters from Victorian pioneers: being a series of papers on the early occupation of the colony, the aborigines, etc*. Melbourne: RS Brain.

Butler, R.W. (1980). The concept of a tourist area cycle of evolution: implications for management of resources. *The Canadian Geographer*, XXIV(1):5-12.

Boyce, J. (2011). *1835: The Founding of Melbourne & the Conquest of Australia*. Collingwood: Black Inc.

Clark, I.D. (1990). *Aboriginal languages and clans: an historical atlas of western and central Victoria, 1800-1900*. Melbourne: Department of Geography and Environmental Science, Monash University.

Clark, I.D. (1995). *Scars in the landscape: A register of massacre sites in western Victoria, 1803-1859*. Canberra: Australian Institute of Aboriginal and Torres Strait Islander Studies.

Clark, I.D. (1998). *Understanding the Enemy: Ngammadjidj Or Foreign Invader?: Aboriginal Perception of Europeans in Nineteenth Century Western Victoria*. Clayton: Faculty of Business and Economics, Monash University.

Clark, I.D. (2002). The ebb and flow of tourism at Lal Lal Falls, Victoria: a tourism history of a sacred Aboriginal site. *Australian Aboriginal Studies*, (2), 45-53.

Clark, I.D. & Heydon, T.G. (2002). *Database of Aboriginal Placenames of Victoria*. Melbourne: Victorian Aboriginal Corporation for Languages.

Clarke, M. (1980). *Big Clarke*. Carlton: Queensberry Hill Press.

Department of Planning and Community Development (DPCD) (2008).*Aboriginal Flaked Stone tools.* Department of Planning and Community Development. Victorian Government. Retrieved January 25, 2013 from http://www.dpcd.vic.gov.au/__data/assets/pdf_file/0011/35876/AA_04_StoneTools_12.06.08.pdf

Department of Planning and Community Development (DPCD) (2012).*Registered Aboriginal Parties.* Department of Planning and Community Development. Victorian Government. Retrieved February 2, 2013 from http://www.dpcd.vic.gov.au/indigenous/aboriginal-heritage-council/registered-aboriginal-parties

Department of Primary Industries (DPI) (2000).*Sugar gum for farm forestry.* Department of Primary Industries. Retrieved January 29, 2013 from http://www.dpi.vic.gov.au/forestry/private-land-forestry/individual-species/sugar-gum-for-farm-forestry

Department of Sustainability and Environment (DSE) (1996). *Midlands Forest Management Plan – Chp 4 Timer production.* Department of Sustainability and Environment. Retrieved January 29, 2013 from http://www.dse.vic.gov.au/forests/publications/plans/forest-management-plans/midlands/timber-production

Department of Sustainability and Environment (DSE) (2012). *Bushfire history–Major bushfires in Victoria.* Department of Sustainability and Environment. Retrieved January 30, 2013 from http://www.dse.vic.gov.au/fire-and-other-emergencies/major-bushfires-in-victoria

Flinders, M. (1814). *A voyage to terra Australis,* 2 Vols. London: G. and W. Nicol.

Gunn, C.A. (1994). *Tourism Planning: Basics, Concepts and Cases.* Washington: Taylor and Francis.

Harcourt, R. (2001). *Southern Invasion-Northern Conquest: Story of the Founding of Melbourne.* Blackburn South: Golden Point Press.

Labilliere, F.P. (1878). *Early history of the colony of Victoria: from its discovery to its establishment as a self-governing province of the British Empire*, 2 Vols. London: Sampson Low, Marston, Searle & Rivington.

MacCannell, D. (1976). *The tourist: A new theory of the leisure class.* New York: Schoken.

McCrae, H. (Ed.) (1934). *Georgiana's journal: Melbourne a hundred years ago.* Sydney: Angus & Robertson Ltd.

Massola, A. (1968) *Aboriginal Place Names of South-east Australia and their meanings.* Melbourne: Lansdowne Press.

Massola, A. (1969). *Journey to Aboriginal Victoria.* Adelaide: Rigby Ltd.

Illustrated Australian News 21/2/1877; 22/1/1879

Norris, R. (2010). *Wurdi Youang, Australia – case study 4.3.* ICOMOS-IAU Thematic Study as part of the UNESCO World Heritage Convention 2010. Retrieved January 29, 2013 from http://www2.astronomicalheritage.org/images/astronomicalheritage.org/thematic-study/ch04cs3.pdf

North Melbourne Courier and West Melbourne Advertiser 4/2/1910

Odgers, L.L. (1857). *You-Yang's summit: a narrative of the proceedings of a pic-nic party: their conversation: with an account of the brilliant scenery, magnificent landscapes, and unparalleled grandeur of the surrounding country in view from the summit of the mount.* Thomas Brown, Geelong.

Parks Victoria (2010). *Parks Victoria annual report 2009/10.*Parks Victoria. Retrieved January 31, 2013 from http://parkweb.vic.gov.au/__data/assets/pdf_file/0010/305011/pvar_200910.pdf

Parks Victoria (2012). *New facilities boost the You Yangs experience.* Parks Victoria. Retrieved January 15, 2012 from http://parkweb.vic.gov.au/about-us/news/new-facilities-boost-the-you-yangs-experience

Parks Victoria (2013). 'Visitor number monitor – 1997 to 2001'. Unpublished report. Parks Victoria.

Pearson, E. (2011). 'Lara's You Yangs 'about to reopen'. *Surf Coast Independent* (24/6/2011). Retrieved February 2, 2013 from http://www.starnewsgroup.com.au/indy/torquay/232/story/135790.html

Pescott, T. (1995). *The You Yangs Range.* Belmont: Yaugher Print.

Roberts, D. (2008). A history of boneseed control in the You Yangs Regional Park, Victoria. *Plant Protection Quarterly*, 23(1):51.

Scott, E. (Ed.). (1966). *Australian Discovery by Land*. Johnson Reprint Corporation.

Scott, E. (2001). *The Life of Matthew Flinders*. Melbourne: Angus & Robertson.

Smyth, R.B. (1878). *The Aborigines of Victoria*, 2 vols. Melbourne: Victorian Government Printer.

Stieglitz, K.R (ed.) (1964) *Emma von Stieglitz: her Port Phillip and Victorian Album*. Hobart: Fullers Bookshop.

The Age 12/5/2006; 7/4/2012

The Argus 22/11/1856; 28/10/1857; 30/12/1857; 7/11/1867; 26/10/1868; 18/6/1890; 25/11/1924; 23/10/1936; 19/12/1950

The Bacchus Marsh Express 6/3/1869; 18/4/1896

The Ballarat Star 27/11/1858; 4/9/1869

The Geelong Advertiser 1/12/1851; 26/5/1914; 20/5/1915; 29/5/1915; 2/7/1915; 1/3/2011; 8/8/2011.

The Geelong Advertiser and Intelligencer 17/12/1853

The Maitland Mercury & Hunter River General Advertiser 30/10/1856; 27/7/1867

The Sydney Herald 1/10/1840

Todd, W. Journal June-November 1835, Typescript copy 16 pp, State Library of Victoria, Ms. 11355.

Victorian Environmental Assessment Council (2006). *River Red Gum Forests Investigation*. VEAC.

Victorian Naturalist (1916) Report of an excursion to the You Yangs ... *Victorian Naturalist*, Vol. 23, No. 395, p. 89.

Waarden, N. V. (1986). Archaeological Survey of the You Yangs, Victoria, *An Occasional report* (Victoria. Dept. of Health and Community Services. Aboriginal Affairs Division), 80.

Webb, C. & Marshall, B. (2002). *An Archaeological Survey: You Yangs regional Park*. Unpublished manuscript. Parks Victoria.

Werribee Shire Banner 23/7/1925

Wynd, I. (1901) *So Fine a Country: a history of the Shire of Corio*. North Geelong. Shire of Corio.

Fig. 9.1: Peak and Valley on the You Yangs (*Illustrated Australian News* 21/2/1877) wood engraving after a painting by JW Curtis.

Fig. 9.2: Holiday Excursionists on the You Yangs (*Illustrated Australian News* 22/1/1879).

The engraving shows men and women enjoying a picnic and views on the summit of Station Peak which includes a wooden lookout tower.

Fig. 9.3: View of Station Peak–(Native You-Yangs)/ taken from the Point Henry Road/ Geelong. / Port Phillip. Gilbert, George Alexander (1856). State Library of Victoria Pictures Collection

Fig. 9.4: Calvert, Samuel (1885). *SCENE IN THE YOU YANGS. Illustrated Australian News* 2/9/1885

Index

A

B

C

D

L

M

N

O

P

R

S

T

U

V

W

Y

Z

Bei Fragen zur Produktsicherheit wenden Sie sich bitte an:
If you have any questions regarding product safety,
please contact:

Walter de Gruyter GmbH
Genthiner Straße 13
10785 Berlin
productsafety@degruyterbrill.com